FUTUREVIEW

Gaining Perspective on the Rising Waves of Change

Mike Woodruff

Christ Church
100 North Waukegan Road, Lake Forest, Illinois 60045
847-234-1001 | www.christchurchil.org

FUTUREVIEW

Gaining Perspective on the Rising Waves of Change

The material contained herein was produced to supplement the FUTUREYOU sermon series held at Christ Church and was intended for primary use by our attenders and other friends of the church. Every effort has been made to fully and accurately ascribe all information derived from others to appropriate sources; if any such errors remain they are inadvertent, and if identified, should be brought to the attention of the church by email to communications@cclf.org, or the postal address listed above.

Cover illustration generously contributed by Guy Billout, a highly acclaimed and award-winning illustrator born in Decize, France in 1941. His style is known for its elegant draftsmanship and visual irony; his full-page feature illustrations have graced the pages of *The Atlantic* (regular contributor, 1982-2006, with complete editorial freedom), as well as at *The New Yorker* (regular contributor, 2001-2011). His work has also appeared in *Business Week, Esquire, Fortune, The Los Angeles Times, New York Magazine, The New York Times, Oprah, Time Magazine, Travel & Leisure, Vogue, The Wall Street Journal,* and *The Washington Post.* Billout got his start as a graphic designer in Paris before moving to New York in 1969. In addition to his magazine illustration, he has also written and illustrated nine children's books, five of which received awards from *The New York Times* for being among the year's Best Illustrated Children's Books (for 1973, 1979, 1981, 1982, and 2007). He has been teaching at Parsons School of Design in New York since 1985. You can see many of his illustrations for *The Atlantic* at *TheAtlantic.com/author/guy-billout/* or visit his website at *GuyBillout.com.*

ISBN 978 - 0 - 692 - 75054 - 4

Contents

Acknowledgments

These are the numbers... who came to David...: from Issachar, men who understood the times and knew what Israel should do...

– I Chronicles 12:32

As with all writing projects, there are many to thank. Let me start with The Murdock Charitable Trust, which funded the gathering that led to this book, and also the Hindustan Bible Institute, which allowed me a quiet place to write in the early days of my research. More than twenty people read various drafts and offered helpful comments. If this work is helpful to you, they deserve much of the credit. Where there are mistakes—which any book that attempts to predict tomorrow will have—they are mine. Special thanks to the staff of Christ Church (a talented and dedicated group that put up with me being more distracted than usual) and to all of the people of Christ Church—which is a great congregation to serve—and from them, the unnamed editor-book designer pair that helped me carry the ball the excruciating last five yards. And finally, special, special thanks to my family, especially Sheri, who humored and tolerated me while I processed my thinking about the future out loud.

Why Study the Future? 1

It's tough to make predictions, especially about the future.

– Yogi Berra

When I was a kid I was told that someday I'd fly a helicopter to work. There were even pictures in a *Popular Mechanics* magazine my third grade teacher kept at her desk. I couldn't wait.

Alas, I'm still waiting.

For the record, it's not just the helicopter that hasn't arrived. Other promises were made. *The Atlantic* said that by the year 2000 we'd have "abolished war and the poor would be living in high-rise 'abodes of happiness and health.'" The chairman of the U.S. Atomic Energy Commission said nuclear power would be "too cheap to meter." And the writers of *Ladies Home Journal* said that all rats and mice would be eliminated, as would the letters C, Q and Z.[1]

As I understood it, by the time I was thirty we'd all be vacationing on the moon, celebrating the end of cancer, breathing clean air, and living to be three hundred (while looking, acting, and feeling as if we were twenty-five). We'd also all have plenty of money and little to do other than jet around enjoying our toys.

But somewhere along the way things took a proverbial left turn. As the cover of a recent issue of MIT's *Technology Review* stated, "You promised me Mars colonies. Instead I got Facebook."[2]

Predicting the Future is Hard

Yesterday's errant predictions shouldn't be much of a surprise. When you study "the history of the future" you see that we've never been very good at looking ahead. We predict things that do not happen, fail to predict things that do, and sometimes get it exactly backwards. Robert Malthus is Exhibit A for the last category. In the nineteenth century he warned that because our population was growing exponentially while our food production was not, we'd all soon starve.[3]

Of course we haven't. Just the opposite. Farmers now grow more food on less land. And the rest of us have gained weight.

And Malthus's misfire is just one of many. Back in the 1970s the CIA failed to foresee Iran's embrace of radical Islam. More recently they were surprised by both the Internet and the Arab Spring.[4] Come to think of it, no one did a particularly good job predicting social media, Lasik surgery, Starbucks, or salad in a bag.[5]

But that should not slow us down. In fact, we desperately need to look ahead, and now more than ever.

Why Study the Future

Why do I say this? Why should you look ahead when futurists have been so wrong so often? Why read a book about the next thirty years when no one can consistently predict tomorrow's weather?

Though we get a lot wrong, we also get a lot right. Yes, there are celebrated misses, but we are getting better all the time. Futurists do not do details very well. But they are increasingly good at identifying the trends that matter.[6]

Looking ahead helps us prepare. This book will help you navigate the emotional and practical trauma associated with change by buying you a bit of time to prepare.

Perhaps you are energized by change. If so, give yourself a high five because you are unique. Most of us find it at least a little unsettling, and it causes some to break out in a cold sweat. "Stop everything! I can't keep up. The voice-activated internet of things? Gene editing in personalized medicine? I don't even know what these things are, let alone how to use them. I still can't program my VCR and they stopped making videocassettes fifteen years ago. I want off the change train."

This response is normal. When books were introduced, people feared them, just as they had feared the rise of writing centuries earlier.

"Books? People feared books? What's to fear? Paper cuts?" You might think those overwhelmed by books were overreacting. After all, words on a page are not exactly disruptive technology." But those yelling "stop" saw it differently. They worried that this new method of communicating would take ideas too far too fast. They reasoned that knowledge had always been linked with memorization and argued that if books became common no one would bother to memorize anything anymore.[7]

And yet, can you imagine where we'd be without books?

I am not suggesting that all change is good. Some is and some is not. I'm simply contending that one of the reasons we need to look ahead is to prepare ourselves for the "books" that are coming our way. Many of us do better with a bit of lead time. I sure do. Sometimes ten minutes is all I need to quell the panic and turn the problem into an opportunity. But I need those ten

Our VUCA World.

–

One of the ways that people are framing the world today is with the acronym "VUCA." As presented by management professor Paul Kissinger:

V = Volatility, the nature, magnitude, and dynamics of change;

U = Uncertainty, the growing unpredictability of events and issues;

C = Complexity, the chaos in which organizations find themselves, and the ways in which issues are confounded; and

A = Ambiguity, the "haziness of reality and the mixed meanings of conditions."[9]

In many ways tomorrow will be like today. In other ways it will be different. But even when it will be like today, it will be faster.

minutes (or ten months). Looking ahead helps us buy time, perhaps enough to keep our wits about us.[8]

We cannot not look. A third reason to consciously look ahead is to better leverage what we are already doing anyway. Face it, you have views about tomorrow. In fact, you have plans for next year. We are all futurists, and our view of the future is shaping the way we live today. The question is not whether we have thoughts about tomorrow. The question is, are our thoughts about tomorrow any good?

Tomorrow will be different. Finally, a major premise of this book is that tomorrow will be different and it's foolish not to prepare.

This is one of the main reasons I am writing this book. We are all going to need a strategy for navigating currents that are becoming increasingly powerful.

Though change is constant, there are times when it is more aggressive than others. The Agricultural Revolution was one. The Industrial Revolution was another. Some argue that we just passed through an Electronic Revolution and have now entered The Fourth Industrial Revolution.[10]

Whether this is true or not, the pace is accelerating. There was a time not long ago when fathers taught their sons the family craft. The lessons they passed down were virtually identical to the ones their fathers had passed down to them. Little changed over the years. We are not living during one of those periods. Today we have to look ahead just to stay in the same place. And as fast as today is moving, in many ways tomorrow will be faster.

I am not writing as a futurist. I am a guy who started reading about the future and didn't stop. In this small book my goal is to provide you with a flyover of the field.

Let me say it again, we are going to need a strategy going forward. Trying to stay on top of things by working harder and longer hasn't been a good idea for a while. It is becoming a worse idea by the day.

We Are Living in a Dynamic Moment.

For most of history, Rip Van Winkle could take his twenty-year nap and not miss much. But that is not true today. Can you imagine showing up for work tomorrow after a twenty-year nap?

What if you'd been out for fifty years? Can you imagine how disoriented people would be if we plucked them, say, off a Detroit assembly line back in the mid-60s, and dropped them in Los Angeles today? What would shock them the most: the collapse of the Soviet Union? Lady Gaga? Bottled water? WiFi?

From Sabbath to Sunday

As a pastor I have come to believe that many people move too quickly for their own good. It's the classic "improved means to diminished ends" problem. One of the reasons is we've lost meaningful downtime.

One hundred years ago we engaged in Sabbath rest every seven days. The Sabbath was structured for reflection and restoration. About fifty years ago we traded it in on Sunday, which became a day to relax.

Relaxing is very different than reflecting. In the first we watch football, take a nap, and read the paper. In the second we review the week, reflect on life, and think about how to be a better person going forward.

Ten years ago we made another trade. We turned Sunday in for another twenty-four hours of weekend. Now, instead of not thinking about work, we do our best to catch up.

We made a very bad trade.

I believe the future is going to move faster than the present and those who try to navigate the accelerating pace by moving faster yet are not going to make it. I do not have three steps to solve this problem or to help you get everything under control. But I do think I can start a conversation that will help you prepare for the pace and change racing your way.

The Four Bigs

This book pivots around four trends shaping tomorrow: the *Acceleration of Technology; Changing Social, Sexual, and Marital Dynamics; Globalization;* and *Swirling Ideologies and Religious Beliefs.*

In the book I refer to them as "Glaciers," in light of a conversation I had with University of Chicago historian Martin Marty.

About a month after the 9/11 attack, I had arranged the interview with Marty—who is the author of over fifty books and the recipient of more honorary doctorates than any other living person—intending to ask him questions about trends in higher education. But given the national mood in October of 2001 I asked him about the impact the attacks would have long-term. He rightly said, "Not much." He then went on to argue that history is shaped by two kinds of events: earthquakes and glaciers. Earthquakes are dramatic and consequently generate lots of press. But as soon as the earth stops shaking, people start putting things back the way they were. As a result, earthquakes do not result in lasting change. Glaciers, on the other hand, move so slowly that we stop paying attention to them. However, once they carve up an area that region will never be the same.

In light of this, Marty said, "We need to focus on glaciers not earthquakes."

That is what we will do.

An Overview

The outline of the book is simple.

You're reading chapter one. In chapter two, *Many Things Are Getting Better,* we'll explore the ways life is improving. I want us to start here not only because I believe many things are trending well, but because I think we overlook much of it and because we need a counterbalance to the pessimism that gets all the press.

Let me be candid, there are some storm clouds on the horizon and it would be irresponsible to ignore them.

- The Doomsday Clock was recently advanced to three minutes until midnight—a position as dire as the darkest days of the Cold War.
- Cambridge physicist Stephen Hawking believes that we will blow up the planet in the next twenty years.
- Elon Musk is losing sleep over Google's experiments in Artificial Intelligence, and management guru Jim Collins (the author of *Good to Great* and *Built to Last*) is now studying what separates "those who do well from those who do not when the world spins completely out of control," because he believes that:

  > There will be no New Normal. We are now dealing with a world that is going to be ferocious. The volatilities, the turbulence, the uncertainties of the world will probably define the second half of my life.[11]

Every field has a lunatic fringe and future studies has a large one. But I would not put any of these people in that camp. I share some of their concerns. However, they will not reflect the tone of the book. For starters, I do not look ahead with fear.

Why not? How can I write about a world of "volatilities, turbulence, and uncertainty" and sleep through the night? There are several reasons.

- I've read enough history to know that humanity has survived wars, plagues, riots, revolutions, famines, pandemics, floods, and dust bowls. We are more resilient than many give us credit for.[12]
- People have been predicting disaster and calamity for thousands of years (Y2K anyone?). The threats ahead of us are real, but good people are working on solutions. I believe that many of them will be avoided.[13]

» I view the future through the lens of faith. I am not saying that I'm naturally optimistic or implying an unbounded confidence in humanity. I'm stating that I believe in a good God who shapes—and in the end, ultimately controls—all things. My experience is that confidence in God provides stability and peace in the face of some challenging forecasts.

In chapter three I will explain my approach and explore my "Monsters Under the Bed," and in chapters four through eight we will look at each of the glaciers, noting what they are, where they are headed and how they will affect us. Finally, in chapter nine—which I consider the most important part of the book—we'll consider how we live in light of where we are headed.

My hope is that this small book will start some conversations that prepare you to navigate the changes racing our way.

A note on Notes.

–

I set out to write a short book. That has forced me to dance lightly over subjects that could be explored at much greater length. If you want to go deeper you can turn to the endnotes section that appears in an appendix, where I not only cite sources, but also develop some of the arguments at greater length.

Many Things Are Getting Better 2

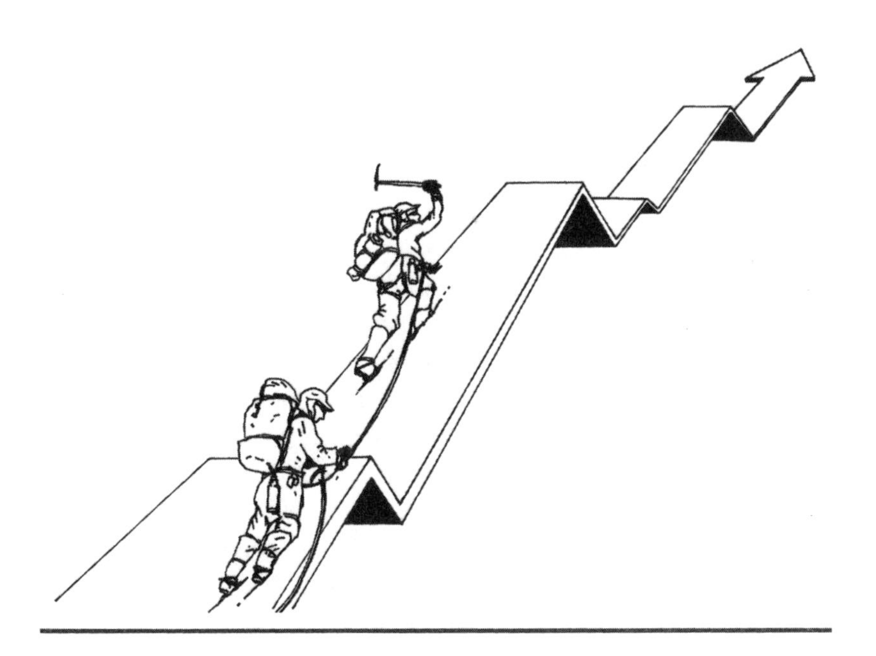

The good news about bad news is that there is not nearly as much of it as you might think. The bad news about good news is that good news doesn't tend to sell.

– John Ortberg

If its individual citizens, to a man, are to be believed, [America] always is depressed, and always is stagnated, and always is at an alarming crisis, and never was otherwise...

– Charles Dickens

The first thing I asked Kevin Kelly, founding editor of *Wired* magazine and author of several books on the future, when I interviewed him for this book was, "Are you optimistic or pessimistic about the future?"

"I am wildly optimistic," he replied. "How about you?"

"Me? Well, I'm more optimistic now than when I started my research. But I am scared about a few things," I said.

"As you should be. What are you scared about?"

"Jobs for starters. I am worried that technology is going to replace lots of jobs, which is going to set a number of bad things in motion."

"Don't worry about that," Kelly responded. "Technology will replace lots of jobs. And it's going to change the jobs it doesn't replace. But it will create lots of jobs as well. It may even create more jobs than it replaces. If you'd told a farmer two hundred years ago that ninety percent of the farming jobs would be eliminated by technology, he might have panicked. But that's because he didn't know about all the new jobs that would be created—web designers, tax accountants, and things like that."[14]

That led us into a conversation that went on for another hour.[15] I'll reference Kelly again. Right now I simply want to note that he sees good things ahead. And that makes his views uncommon.

If you ask five random people if they think we are heading in the right direction or not, you usually hear that we are not.

Utopian views were common at the start of the twentieth century. But the Great Depression, two World Wars, the Cold War, nuclear proliferation, Vietnam, climate change, 9/11, suicide bombers, school violence, and the like have left people jittery.

In 1979 the university I was attending almost canceled a debate on the question, "Are Things Getting Better or Worse?" because they couldn't find anyone willing to defend the idea that things were getting better.[16]

Around that same time, Julian Simon, professor at Maryland, grew so frustrated with the gloom and doom forecasts coming from the popular Stanford professor, Paul Ehrlich, that he demanded that Ehrlich "put up or shut up."[17] Simon was an economist who believed things were improving. Ehrlich was a professor of population studies who said they were not. In fact, Ehrlich claimed that within a few years we were all going to starve. And he was not only saying as much in his lectures and books, but also on the Tonight Show and in movies.

In an effort to silence Ehrlich, Simon issued a public challenge. He invited the Stanford professor to pick the five commodities he thought we would run out of first. Simon was willing to wager that in ten years all five would be cheaper and more plentiful.

Ehrlich took the bet and lost. All five went down.[18]

Think about that for a moment. Ehrlich claimed that everything was out of control and we were all about to die. Simon called his bluff. He allowed Ehrlich to control the bet (i.e., to tilt things in his favor by selecting the five areas that were certain to be the most problematic). Ehrlich did. He controlled the variables and yet he still lost the bet.

Before We Look Ahead We Need to Look Around

Before we discuss tomorrow, we need to agree on today. I suspect that it is better than you think. Of course not for everyone. There are many who are struggling. But on aggregate the trend line has been positive.

I say this after spending a year chasing facts and pouring over statistics. The media serves up a diet of "Cancer on the Rise!" or "Zika to Decimate Africa." Or, "Worst Flooding in Years" (or "Worst Drought..."). Worst This and Worst That. Our sense from all we hear is that crime is escalating, our children are falling behind, and we will be out of topsoil by noon tomorrow.

But the truth is, many things are trending well.

» In 1820, average worldwide life expectancy was 29. Nearly 200 years later it is 70 and climbing.[19]

» 150 years ago, 75 percent of the world lived in extreme poverty; today it's down to 10 percent and is continuing to decline.[20]

» Pollution, crime and illegal drug use are all down. Access to clean water is up.[21]

» Global literacy rates have climbed from 21 percent in 1900 to 83 percent today.[22]

And on average, in the United States, we now start working two years later and retire sooner than we did in 1950. Paid holidays have doubled and time off for vacation is up by 63 percent. Working conditions are better and we have more benefits and better tools. Many of today's poorest households have more material goods (e.g., a dishwasher, freezer, microwave, color TV, air conditioner, etc.) than the average household had twenty years ago.[23]

Sociologists Fischer and Hout, on the Last Century.

–

"The twentieth century was, with the notable exception of the 1930s, one of prodigious economic advancement in America. While working fewer hours, Americans easily quadrupled their real earnings."[26]

–

Claude Fischer and Michael Hout

If you ask the average person, he will say that he is working harder and longer than ever before. But it's not true.

In fact, it's not even accurate to say we live like kings of old, because most of us live much better than that. The average supermarket in the West offers thirty thousand items—an array of choices well beyond the reach of royalty a century ago.[24]

Surprised? There is more. High school dropout rates, college enrollment, juvenile crime, drunken driving, traffic deaths, infant mortality, workplace injuries, air pollution, divorce, charitable giving, voter turnout, per capita GDP, and teen pregnancy are all improving. We are better educated and more literate than in the past. We have less crime, fewer people die in war and more live under democracy.[25]

And these improvements are not limited to the West. In a recent *Newsweek* column, Fareed Zakaria noted the improvements being made all over the world.

> The world's tallest building is in Dubai. Its largest publicly traded company is in Beijing. Its biggest refinery is in India. Its largest passenger airplane is built in Europe. The largest investment fund is in Abu Dhabi. The biggest movie indus-

try is Bollywood. The largest Ferris wheel is in Singapore.[27]

Some see Zakaria's list as a collection of America's failures (i.e., we are falling behind). But Zakaria sees this as a partial list of the ways things are improving around the globe. I think he is right.

I am not implying that we do not have problems. We do. Chicago's poor neighborhoods remain riddled with gang violence. Two of the communities close to where I live are food deserts. And NGOs are right when they claim that one child dying from lack of clean water is one child too many.

But there's little to be gained by crying wolf when there is no wolf around, or by overlooking all the things that are going well. We think crime is getting worse but it's been dropping for some time now. We think we are working more and playing less but the opposite is true. As the figures below show, many things are trending well. In fact, for some time now, most things have gotten better for most people.[28]

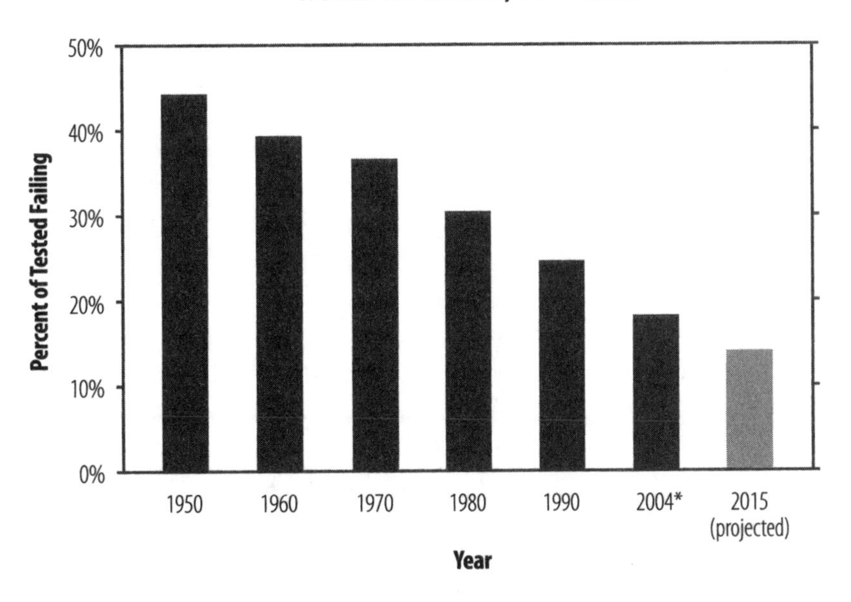

Figure 2.1 The drop in global illiteracy over the last sixty-five years, a drop of approximately two-thirds the 1950 level (2005 analysis, UNESCO projection of 86% literacy by 2015).[29] Note, asterisk is used to indicate that data as 2004 were gathered over the period of 2001-2004.

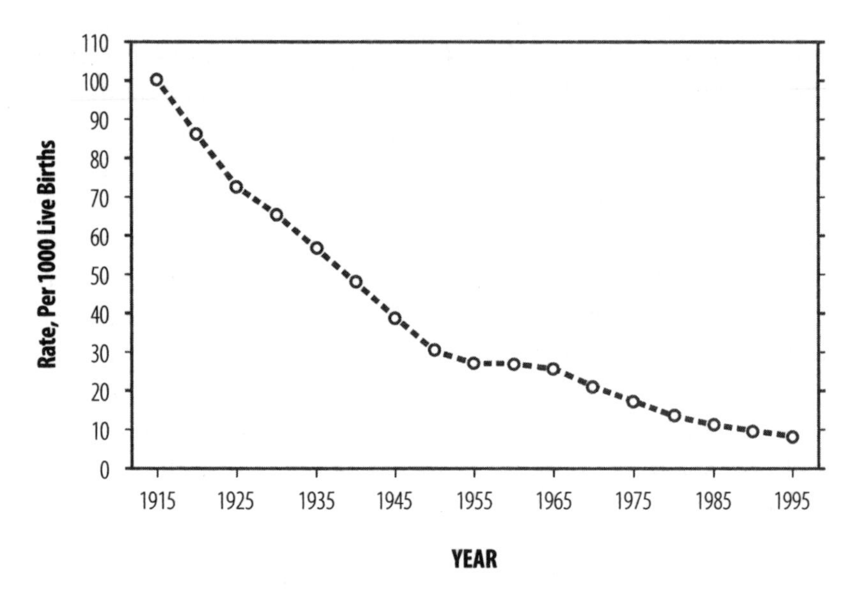

Figure 2.2 The drop in U.S. infant mortality over the 20th century, a drop by 1995 of approximately 90% of the 1915 level (CDC analysis).[30]

I could fill the next twenty pages with figures that are similarly remarkable.

Why No One Tells You Things Are Getting Better

"Wait," you say. "If so many things are improving, why aren't more people talking about it?" It turns out there are a number of reasons.

Good news doesn't sell papers. Which story gets your attention: "Local Celebrities Donate to Homeless Shelter," or "County Exec Caught Embezzling!" Or "Tap Water Meets Standards," or "We Are Poisoning Our Children!?" *And should it have?* Bad has more cachet than good. The more provocative, alarming, and frightening a story is, the more likely we are to read it.[31]

And getting you to read the story is the goal. After all, the news business is a business. Editors, publishers, website hosts, and other journalists need your attention to drive ad rates, which drive the revenue they need to survive. More eyeballs equal more money. Given that fear sells, they avoid stories about clean rivers and stable economies to alert us that a comet is going to slam into earth, never mind that it's not due for three-hundred thousand years. Bad news is favored.

The editors are not stupid. They know that Paul Revere didn't get famous for saying, "The British stayed home! Go ahead and sleep in."

Some people mislead. Some bad news is a matter of marketing, some is bad reporting, and some is worse than that. Remember the story claiming that in the 1940s schoolteachers complained about children "talking, chewing gum, and running in the hall"? Whereas now they complain about children abusing drugs, getting pregnant, and committing assault?

It's an amazing story. The problem is, it isn't true. A reporter made it up.[32]

And it didn't just happen once. Years ago a *Pravda* headline read, "Soviet Runner Second, U.S. Third from Last!" While technically true, the editor left out an important detail: there were only three people in the race.

Once you learn how to spin, you can "adjust" just about any story. Increasing life-expectancies become yet another reason our Social Security system is going to fail. A drop in gas prices is turned into a report about declining profits for oil companies.

Because good news has more grab, we get bad news more often and on more channels. And sometimes in hi-def.

Others Besides News Agencies Are Also Channeling Eeyore.

–

Those raising money for a cause are also motivated to make things sound worse than they are.[33] Why? Because if things are dire, we will send more money. A while ago, Greenpeace mistakenly posted a draft of a press release that read:

In the twenty years since the Chernobyl tragedy, the world's worst nuclear accident, there have been nearly [FILL IN ALARMIST AND ARMAGEDDONIST FACTOID HERE].[34]

We Have a "Bad Bias"

When people are asked if things are getting better or worse, 83 percent say worse.[35] This sounds compelling until you realize that we've been reporting "worse" for the last forty years—during which time many things have gotten better.

In a 2003 study designed to measure the accuracy of our perception of children's well-being, three-quarters of those surveyed thought the number of children on welfare had gone up when it had gone down; likewise, in a similar study, 90 percent thought crime rates among teens had gone up when they were at a twenty-five-year low.[36] These studies suggest that we believe things are worse even when they are demonstrably better.

Numerous other studies show that we believe things are spiraling apart even when we are doing better. David Whitman calls this phenomenon the "optimism gap." Sociologist Bradley R.E. Wright describes the phenomenon of the "grass is browner, not greener" (on the other side of the fence), noting that it is referred to by psychologist Fathali Moghadam as "the sky is falling, but not on me," and by journalist David Whitman as the "optimism gap."[37]

Why do people think this way?[38] Perhaps we've learned to keep our expectations in check so we are not disappointed later. Maybe we have unrealistic expectations about life (i.e., we are not happy with progress because we expect perfection). I do not know. Whatever the cause, we need to be aware of our pessimism bias.

There Is Plenty of Bad News That Is Real

Before we swing too far in this direction, let me again acknowledge that there are some troubling trend lines.

In his 2011 book, *Upsides: Surprising Good News about the State of Our World,* sociologist Brad Wright expresses frustration over the pervasiveness of negativity and attempts to document the high cost our pessimism brings.[39] But he also notes that we do face real problems. My summary:

> There is too much debt and too little savings; we have more free time but we tend to spend it watching TV; air and water quality is up but the climate is warming; abortion is down, but on most other categories involving social relationships, the news is bad: marriage and two-parent families are less frequent, cohabitation (which leads to less

stable marriages in the future) is up. And there is obesity, social isolation, and more.[40]

There are things to be concerned about.

Somewhere between calling bad *good* and crying wolf when no wolf exists is the right response. That is what we are after.

It would have been easy to start this book with scary statistics and doomsday scenarios. But I do not think that would be accurate, and I want to lean towards the positive whenever I can.

»«

In the next chapter I will introduce the glaciers reshaping our world, and the "Monsters Under the Bed" that could disrupt everything.

3 Glaciers Reshaping the Globe

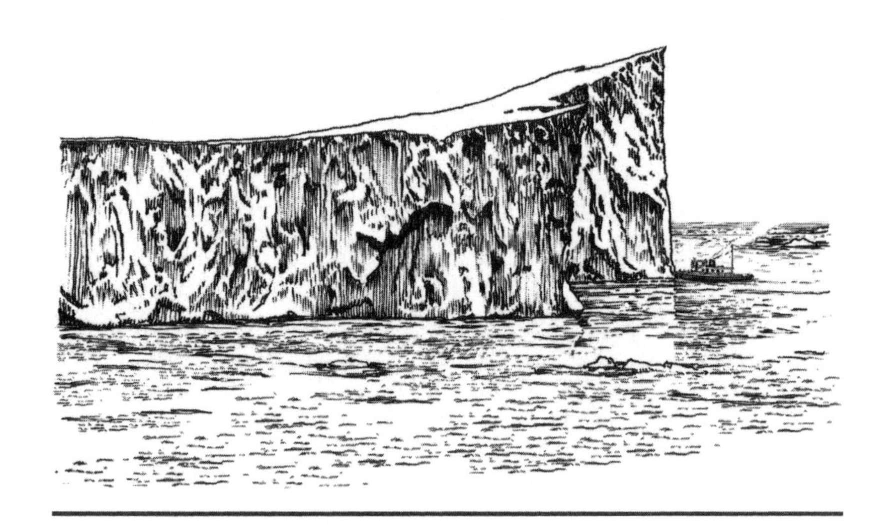

Whirl is king, having driven out Zeus.

– Aristophanes

These are the numbers... who came to David....: from Issachar, men who understood the times and knew what Israel should do...

– I Chronicles 12:32

*Skate to where the puck is going,
not where it is.*

– Wayne Gretzky

Science and technology drive change. So do environmental issues, immigration, religion, politics, war, the economy, and a dozen other things, including pain. Chaos theory goes so far as to suggest that a butterfly flapping its wings in Brazil might change the weather in Boston.

Whether the "butterfly effect" is valid or not is for others to debate. We are going to focus on the four biggest trends affecting the future. While acknowledging that everything affects everything, we will drill down on those items that are making the biggest difference: the *Acceleration of Technology; Changing Social, Sexual, and Marital Dynamics; Globalization*; and *Swirling Ideologies and Religious Beliefs.*

For what it's worth, I am hardly alone in thinking the best way to look into the future is to focus on the macro-trends. In *The Meaning of the Twenty-first Century,* James Martin discusses what he calls "trends of unstoppable momentum that are forming a skeleton of our future." In his best-selling book of the '80s, John Naisbitt explores "megatrends."[41] And in a report written for the McKinsey Global Institute, Richard Dobbs, James Manykia, and Jonathan Woetzel follow the same pattern. The introduction to *No Ordinary Disruption: The Four Forces Breaking All the Trends* includes the following:

> In the Industrial Revolution of the late eighteenth and early nineteenth centuries, one new force changed everything. Today our world is undergoing an even more dramatic transition due to the confluence of four fundamental disruptive forces—any of which would rank among the greatest changes the global economy has ever seen.
>
> Compared with the Industrial Revolution, we estimate that this change is happening ten times faster and at three hundred times the scale, or roughly three thousand times the impact. Although we all know that these disruptions are happening, most of us fail to comprehend their full magnitude and the second- and third-order effects that will result. Much as waves can amplify one another, these trends are gaining strength, magnitude, and influence as they interact with, coincide with, and feed upon one an-

> other. Together... fundamental disruptive trends are producing monumental change.[42]

Other authors have other lists, but you get the point. The best way to look ahead is to map the trajectory of the major trends. We are about to do that. But before we do, I am calling a brief time-out to identify a secondary set of factors. Unlike the trends to which we will devote most of our attention, these events are unlikely. I am listing them here because if any of them do occur, they will have a large impact. And though they could be either good or bad,[43] I call them "Monsters Under the Bed."

To be honest, it seems alarmist to mention them at all. But I decided it would be irresponsible not to. If a terrorist sets off a dirty bomb in Manhattan, if Iran nukes Israel, or if the worst climate change forecasts are true, the future takes a quick left turn.[44]

Time Out for Monsters

Years ago I started collecting articles about world-ending scenarios. I viewed the file as a joke, labeled it "Monsters Under the Bed", and went years at a time without even thinking about it. But at some point I noticed that it had filled up, and as I paged through it I realized that some of the "game-over" scenarios were legitimate concerns.

A good number of the items listed below come from lectures given by Bill McGuire, an emeritus professor at University College London and the Director of their Aon Benfield Hazard Research Center. Prof. McGuire, AKA the "Professor of Doom and Disaster," is focused on events so large that they could wipe out a country or upend the global economy.[45] (His lectures are fascinating, although I doubt he gets many dinner party invitations.)

I picked up other "Monsters" here and there. For simplicity sake, I have sorted them into five groups:

Epic natural disasters. McGuire claims that under certain geological conditions, tsunamis as large as one mile high could sweep across the Pacific, causing hundreds of millions of deaths. He also frets about asteroid strikes, super volcanoes, and algae blooms that choke the oceans.[46,47]

Pandemics. In recent years, the movie industry has stoked fears of deadly viruses spreading around the globe via international travel or terrorism. In some cases they are natural pathogens that have mutated. In others, they have been altered by man. According to experts, both options are possible and potentially devastating.[48] Indeed, during the 1918 Spanish flu pandemic, five percent of the world's population died. Surveillance, and the availability of modern vaccines and drugs have kept outbreaks in check. And it may do so on into the future, but globalization, urbanization and encroachment on natural habitats, and humankind's mobility all increase our susceptibility to pandemics.[49]

Climate change. If the worst-case climate change predictions are true, there is no need to debate the future of New Orleans—or Manhattan for that matter. We should save our energy to develop ocean front property in Ohio. Some books about the future devote a lot of the attention to climate change, believing it will significantly shape tomorrow. I do not say much about it because I do not understand the science well enough to form an opinion. But you should know that if our weather becomes more extreme, this could lead to droughts, to changes in the distribution of animals and disease across the globe, to political instability, and to ecosystem collapse.[50]

Nuclear war. Beyond the saber rattling of Kim Jong-il and the on-again-off-again nature of Iran's centrifuges, lies a serious nuclear threat. Significant stockpiles of weapons of mass destruction are scattered around the globe, and not all of them are accounted for. This is why the Bulletin of the Atomic Scientists moved the Doomsday Clock ahead to five minutes to midnight.[51] One nuclear bomb can ruin your whole day.

The snowball effect. Finally, though each of the four scenarios might happen on their own, some scientists think a snowball effect of multiple events is more likely. For instance: global warming leads to a spike in pathogens in the environment, which makes it harder to produce food, which impacts food production, which leads to escalating prices, which causes political unrest, which leads to terrorism, which eventually includes nuclear weapons, etc.[52]

Under this scenario, no epic catastrophe undoes us, but a cascading array of small factors work together to degrade life on

Earth. One writer suggests that the downfall of Earth will not be dramatic (i.e., like a asteroid impact or single massive pandemic). Instead, biology professor Joseph Miller says it will be more "like being nibbled to death by ducks."[53]

On to Our Likely Future

As we turn from "Monsters" to "Glaciers," let me remind you that I believe the risk of the "game-over" scenario is low.[54]

I do not live in fear because I believe bad news is over-reported and good people are working hard to provide ways forward. Also, I write from a position of faith in a good and loving God, who I believe will prevail and has secured a wonderful destiny for his children. This does not mean we will not face challenges. We will. Terrorism is part of the new normal, pandemics will break out, and weather patterns may grow extreme.

But panicking is not a good strategy. Let others play Chicken Little. The way forward starts with understanding what is likely to come our way and then preparing for it.

»«

In the next chapter we are going to explore the impact technology is having in our lives. All signs suggest we're in for a wild ride.

The Acceleration of Technology — Part I

We shape our tools and thereafter our tools shape us.

– Marshall McLuhan

If you think change is hard, prepare for extinction.

– Mike Conley

The twenty-first century began with fears that Y2K would crash society, but that did not happen. The opposite did. Technology sprinted ahead, "google" became a verb, and more than one billion people joined Facebook.

And that was just the warm-up.

Phones, search engines, road maps, encyclopedias, rolodexes, and video cameras were folded into small handheld devices more spectacular than anything Q invented for James Bond. And then these gadgets were handed out to nearly everyone on the planet. (In India the joke is, "Not everyone has a toilet, but everyone has a cell phone.")

Think about the words you learned in the last few years: Twitter, Skype, iPad, Bitcoin, ApplePay, Hulu and SnapChat. Together they suggest that technology is expanding in every direction and doing so more quickly than before.

Whoever you are, you will have access to more power next week than you have today. And you will have even more power the week after that.

There are upsides to this—greater wealth, creature comforts, and cool toys to name a few. But it is disruptive. And it will make life move faster. I do not think we are prepared for what is in the pipeline. We are going to need to learn when to ride the tech wave and when to let it pass us by.

But I am getting ahead of myself. Let me start at the beginning.

Trade-offs.

–

A number of years ago my mom expressed frustration with her dishwasher. When I asked her what was wrong with it, she explained that the dishwasher worked fine. It was its impact she was upset about.

It seems that some of her most meaningful times with her mom came when they stood side by side doing the dishes after supper. A lot of important conversations took place at the sink.

Or perhaps a lot of seemingly unimportant conversations added up to be important. »

Tech 101

What does it mean to live well? What will it look like to live well tomorrow? Let me start by stating a few of my assumptions about technology. These will help us understand the opportunities and challenges heading our way.

One: Technology changes us. The term "technology" (tech) can refer to everything from the spoon you used at breakfast to the Space-X rocket ferrying satellites into orbit. For our purposes I am defining the term this way:

> Technology refers to any item, application or process that helps us save time, multiply power, or otherwise improve our lives, either by making new goods or services possible or by making existing ones easier, faster, or cheaper.[55]

It is common to hear people say that technology is just a tool and that the morality of the device depends on how it is used. That is both naïve and wrong. Technology is moral. It changes who we are by changing how we think and what we do.

Let's set aside dishwashers and air conditioners for a moment and consider the car instead. Ford's Model T not only allowed a person to travel more quickly than they could by horse, its descendants also changed twentieth-century society in a number of other ways. It made people more mobile, facilitated the development of the suburbs, empowered the sexual revolution, grew the middle class, led to an associated rise in obesity, and subtly suggested limits on family size.

Self-driving cars are expected soon and they are expected to have a similar impact.[56]

Because self-driving vehicles will have fewer accidents than human-driven ones, buying car insurance for a human driver will soon become cost prohibitive, and human-driven cars will become as uncommon on city roads as a horse is today.

Whatever the case, looking back she realized that the time she had enjoyed as a daughter with her mother was not being replicated as a mother with her daughters. Kitchen Aid had made her life easier, but now she wondered if the trade-off was worth it.

Other people tell similar stories—before air conditioning, people sat on their porch in the evening to cool down. As a result, they got to know their neighbors. After air conditioning became popular, people stayed inside and neighbors became strangers.

Technology does that a lot. It makes life easier, but it takes something away in the process. And often what it takes away is something we don't realize is important until it is gone.

Soccer moms who double as unpaid chauffeurs will get a break. They can send Jake and Emily to soccer practice on their own.

» Because every car will communicate with every car around it—alerting the other vehicles that in four miles it will be slowing down to turn left—traffic patterns will constantly adjust to maximize traffic flow. This will cut commute times and allow people to move further away from city centers.[57]

» Because cars sit idle ninety-five percent of the time—and depreciate one hundred percent of the time—the fact that they are not great investments will become clearer. In the future, partnerships led by companies such as Uber, Ford, Tesla, and Google will launch fleets of driverless vehicles to chauffer you wherever you want to go. If you are alone, you will summon a small car. If there are six of you, or if you need to haul something, a different kind of vehicle will be sent.

» My point is not that driverless cars are fascinating or that upcoming transportation arrangements will be cool. My point is, technology changes how we live, which changes what we do, how we think, and ultimately who we are. When technology changes, we change.[58]

Two: Technology's impact is expanding and accelerating. Technology is not only changing the way we live, but also the pace at which it is doing so is accelerating.

One thousand years ago, a "new tech release" came along every three hundred years or so.[59] By the 1900s the pace had picked up. In 1950 the rate of change had begun to accelerate, and as Figure 4.1 suggests, new tech releases are now rolling out faster than ever before.

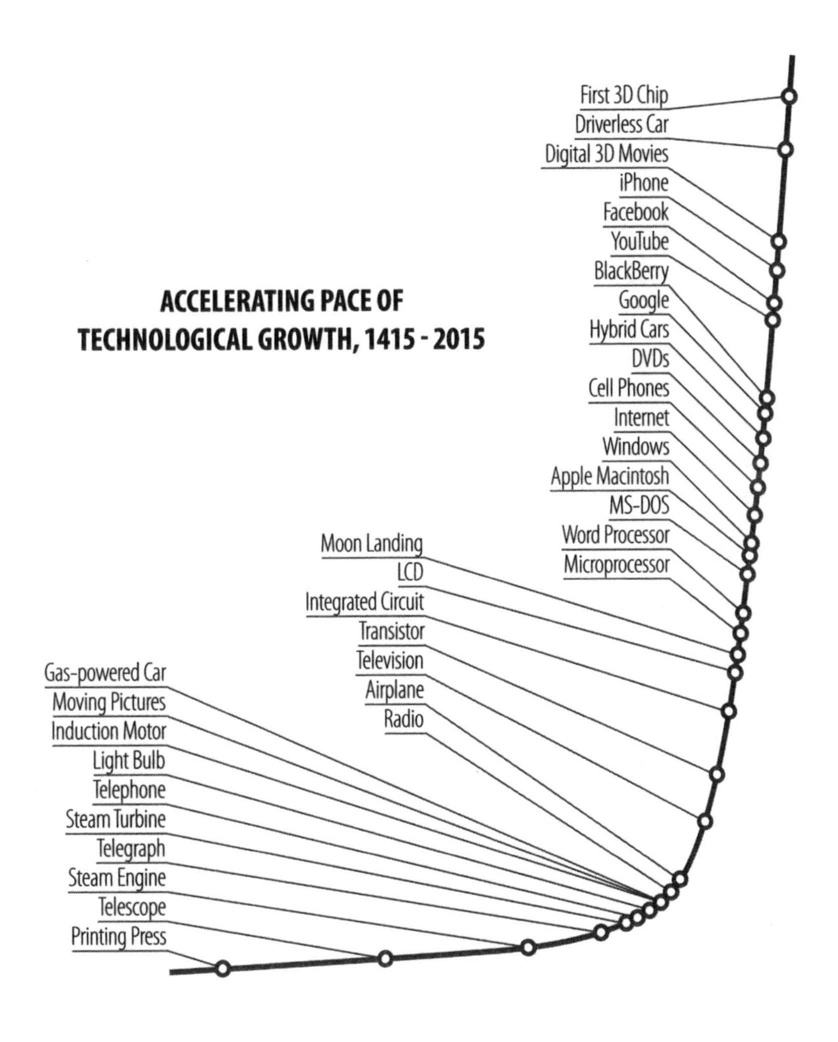

Figure 4.1 The acceleration of emerging technologies in our day, after Kurzweil, Merswolke, and others;[60] a subjective selection of the history of discovery, especially in transportation, communication, and computing.[61]

Given that ninety percent of all the scientists and engineers who have ever lived are alive today, we can expect the pace of technology to continue accelerating. We can also expect to feel overwhelmed, which is what the Tofflers were predicting when they coined the term "Future Shock."[62]

Moore's Law.

–

In the 1970s, Gordon E. Moore, one of the founders of Intel and one of the creators of the integrated circuit board, noted that manufacturers were squeezing twice as many transistors on a circuit every twenty-four months. Given that electrons had less distance to travel, they were able to operate twice as fast. Armed with this insight, Moore predicted that computer processing power would double every twenty-four months. His announcement was met with skepticism, but his predictions proved accurate for the next three decades.[66]

If cars had gained speed at the same rate as computers, they would now travel many times the speed of sound. And if buildings had kept growing taller at the same pace, today's skyscrapers would rise halfway to the moon.[67]

And here is my point: as fast as life is today, it will be faster tomorrow.[63] Ray Kurzweil, the author of The Age of Spiritual Machines (and currently the Director of Engineering and Machine Learning at Google), argues that it's not just that technological change is occurring at an exponential rate. He believes that the rate at which change is accelerating is also increasing exponentially (i.e., we are experiencing exponential change to the nth power). This leads Kurzweil to argue that our linear view of change is dated, that we should prepare for one hundred years of advancement in the next twenty-five years, and for one thousand years of advancement in the following seventy-five.[64]

I think Kurzweil is wrong,[65] but I suspect it may feel as though he is right. We are living in a dynamic moment, and there is little reason to expect that things will slow down soon. Future Shock is present reality.

Three: New waves of change are about to wash over us. Although technology may appear to expand in every direction at once, what actually happens is that a specific technological advance disseminates across multiple "platforms" at the same time. In the first part of the twentieth century it was electricity. It may have felt like everything was changing, but all that was happening was that many different things were being electrified (e.g., clocks, refrigerators, coffee pots, fans, etc.). More recently, massive disruption occurred as computer advances rolled through just about every device and process in existence.

Four: Technology can be misused. So far I have argued that technology changes us and that its impact is expanding. Next I will survey the new waves of change that are about to wash over us. But before I go there, let me share the fourth, and final, of my starting assumptions about technology. It has a dark side. As both history texts and dystopian movies suggest, breakthroughs in science go both ways—they provide us with nuclear power, but they also saddle us with nuclear waste.[68]

The same Internet that helps us navigate unfamiliar streets and reconnect with our best friend from third grade, networks like-minded radicals and provides them with step-by-step instructions on how to build a pipe bomb.

Technology amplifies our character—which means it multiplies good intentions and facilitates evil secrets.

Six Impending Technology Revolutions

One of the reasons many think we will soon be drowning in change is because there are six tech revolutions waiting to wash over us, and several of them possibly as big of an upgrade as electricity was in the early twentieth century. Imagine what it will feel like if even half of them hit at the same time?

What are these revolutions?

» **Nanotechnology.** Small is about to be the new big. And by small I do not mean microscopic. I mean molecular. Nano—which is short for nanometer (a billionth of a meter)—has technologists attempting to use chemistry to build "machines" out of molecules. Imagine if a cancer-destroying "nanobot"—less than one-millionth of an inch—could be injected into your bloodstream on a search-and-destroy mission.[69] As with the virtual reality revolution I will describe, the promises of nanotechnology (nanotech) are running ahead of deliverables, but the hope is that these tiny machines will help us defeat disease, reverse aging, clean up oil spills, and more.[70]

» **Gene-editing and new biologies.** Genetic engineering has long referred to efforts to remove or add genetic material (DNA) to an organism in an effort to change its characteristics. When this is done in the "germline," man-made changes become "heritable." With the advent of recent gene-editing technologies (i.e., the so-called CRISPR methods), alterations are made to facilitate rapid changes to entire organisms. And this is only one of a series of "new biologies" (chemical biology, stem cell biology, synthetic biology, etc.) that are shaking up traditional ideas of the boundaries of science. Some claim that once these technologies are refined the blind will see, the lame will walk, and we'll be able to grow new arms, eyes, or livers if we need them.[71]

» **Additive manufacturing.** Known by its poster-child example, three-dimensional (3D) printing, the emerging field of additive manufacturing (AM) refers to various processes by which an industrial robot synthesizes three-dimensional objects. Some believe AM will launch a third industrial revolution and radically redefine manufacturing in general. They may be right. The space station now has a 3D printer onboard, which allows them to store fewer parts. Remote airports are purchasing 3D printers to produce the parts planes might need, in Dubai a largely 3D printed office building recently opened, and some now claim that by 2030, a full quarter of some urban development efforts with be accomplished using 3D technology.[72]

» **Virtual reality.** The field of virtual reality (VR) provides a computer-simulated, digitally-recorded, 360-degree view of the world. Rather than playing a video game, VR transports you into the game—think Dolby Surround Sound for the eyes.[73] Though over-hyped initially, especially since it gave many people motion sickness, VR is progressing. At a conference I donned a high-end headset and then "walked" through a Syrian refugee camp, while sitting on a stool in the exhibit hall. I'm not as ga-ga about VR as some are. But if the manufacturers fix the things they say they will, golfers will soon be playing Augusta in their living room, football fans will view the Super Bowl from vantage point of their favorite player, and patrons of the arts will tour the Louvre (or attend the Met) without leaving their desks.[74] The educational, occupational, and recreational applications of this technology seem limitless.

» **Big data.** The last category I'll mention here is data analytics, also called "big data," which uses technology to examine large amounts of data in order to uncover hidden patterns, correlations, and other insights. Born in significant part from science's experience with massive data sets (think particle physics) and digital business's desire to track the behavior of customers, analysis of observational data is now being applied to targeted advertising, business process optimization, security issues, financial trading, and even personal health and fitness.[75] At a recent conference I learned that Target has had to rethink the way they apply the insights they gain from their customer's purchasing patterns. In a few instances they realized that a woman was pregnant before her husband had been told and they spoiled the surprise by flooding her with ads for baby diapers. Be prepared for people to combine your Google searches, phone activities, tweets, likes, recent purchases, travel patterns, etc. and know more about you than you may have admitted to yourself.

»«

A little overwhelming? Well, if these astound (or awaken fear),[76] the last of the revolutionary technologies appears poised to surpass them all. Mull on these for a moment, then push on to what many experts believe will be the revolution that changes everything.

5 The Acceleration of Technology – Part II

*The changes are so profound that,
from the perspective of human history,
there has never been a time of
greater promise or potential peril.*

– Klaus Schwab

*Come, let us build ourselves a city,
with a tower that reaches to the heavens,
so that we may make a name for ourselves...
[and] it was called Babel...*

– Genesis 11:4, 9

Alongside the five pending tech revolutions I mentioned—nanotech, gene-editing, VR, AM and big data—is a sixth: Artificial Intelligence (AI). I held it out of the earlier chapter in order to give it more attention here. Why? Because many believe that AI will stand beside the discoveries of fire and the wheel in the museum of Most Amazing Breakthroughs of All Time.

What is AI? You've likely heard about the AI at work in Apple's Siri,[77] and perhaps also AlphaGo, Google's DeepMind project that recently bested South Korea's reigning champion in the game of Go. But what exactly is artificial intelligence and how does it work? Technical definitions describe AI as a sub-field of computer science attempting to enable computers to do things normally done by humans. Further explanations include a distinction between strong AI (machines designed to think like humans) and weak AI (machines designed to provide human-like responses even if they are not produced in the ways a human would think).

But what separates AI from other computing is its ability to learn and adapt. Rather than being programmed to respond with brute computational force—as did IBM's Deep Blue computer when it beat a reigning chess master—AI is designed to think, and then taught how. Most current AI attempts a reverse engineering of the human brain—i.e., uses neural networks—and so uses algorithms with cycles of self-modification that make learning possible. This enables them to master games and other tasks.[78]

This means that AI machines—AlphaGo and beyond[79]—are making their own decisions. Or to make this sound even more ominous, this means AI machines are exercising free will.

When will AI arrive? AI is already here. It's been piloting commercial flights for all but ten minutes for the last twenty years. It's the second set of "eyes" reading your x-rays and it's part of the chip helping your new car brake more effectively.

How quickly will AI spread? We overestimate the impact technology will have in its first few years, but underestimate it after that. At the moment, AI does a decent job answering your email, telling you who the Secretary of Agriculture was during the Eisenhower administration, and updating you on the Cubs game. But it has less common sense than a field mouse.[80]

Improvements to AI are expected to come in fits and starts.

But its promise is enough for many tech investors to claim that by 2026, AI will be a bigger deal than the Internet.

Will my smart phone be smarter than I am? *Is your calculator better at arithmetic than you are? Does your car's GPS system have a better sense of direction than you do? Would you say that either are smarter than you are?* I assume you will concede that even a 1980s Texas Instruments calculator does a better job adding numbers than you do. And a ten-year old Garmin is more trustworthy than you and a Rand McNally atlas. But I am also pretty sure that you do not think they are smarter than you are. Their ability to add numbers or triangulate locations does not equal intelligence.

That approach helps us understand the nature of AI. Hollywood aside, it is never going to be a smarter human than you are. It may be able to "read 800 million pages a second," but it cannot really teach you how to ride a bike or fold a towel.[81]

Let me illustrate this with two figures. The first shows you how not to think about intelligence.

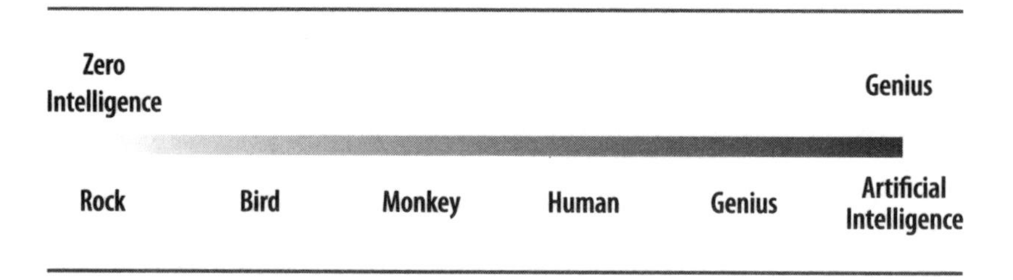

Figure 5.1 AI misunderstood, as a measure of intelligence in one dimension, with more of it in AI devices than humans, in humans than other animals, etc.

It is better to think about intelligence as having multiple dimensions in addition to math-verbal (i.e., there are different kinds of intelligence among people, such as musical, kinesthetic, spatial, emotional, etc.).[82] And in some ways, animals are more intelligent than we are. (For instance, birds that migrate are able to fly thousands of miles and end up in the same spot every year, while some humans get lost a block from their homes.)

Kevin Kelly, whom I mentioned earlier,[83] suggests the following model for intelligence.

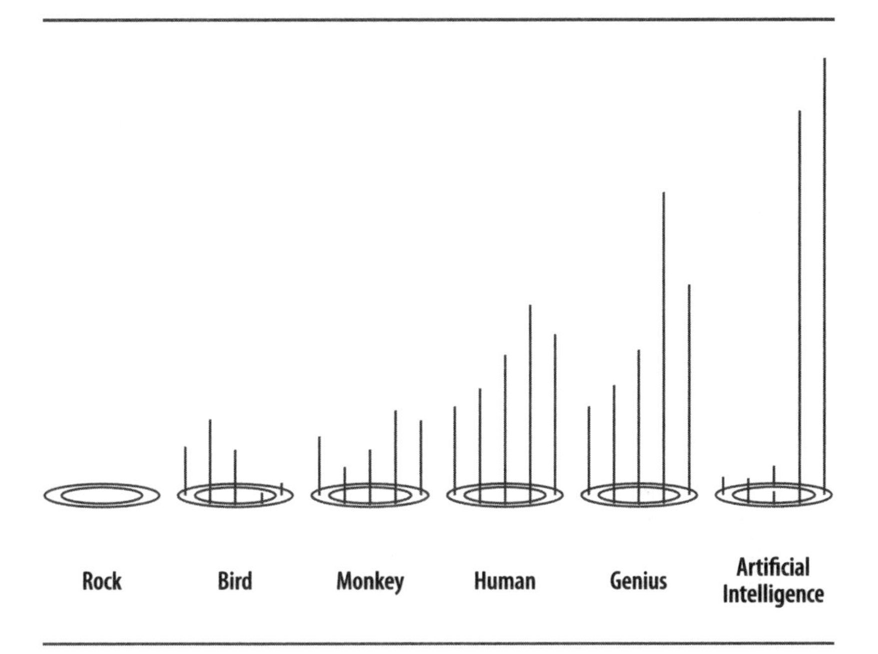

Figure 5.2 AI and multiple intelligences, the view that a subject's intelligence involves varying contributions of different intelligence components (vertical lines).

In other words, AI will be very capable in some realms but clueless in others. As Kelly said to me, "This is exactly why it will be so beneficial. Self-driving cars will be better drivers than humans because the machine will be focused exclusively on driving. Your car will never be distracted by a song or wonder, 'Should I have gone to medical school? Was I supposed to pick up something at the store? Is it going to rain tomorrow?'"

Will AI machines have a conscience? *Will they be self-aware or have personalities, or be alive?* The answer to these questions depends upon whom you ask and how you define certain terms. Many techno-optimists believe that machines will come alive. They contend that an amazing breakthrough will one day rip "the fabric of human history" in half, and that after this happens everything will be different.[84]

Their premise is that we will build a learning machine, AI-1, that will in turn create a second learning machine, AI-2, which will be smarter than AI-1 and able to build AI-3 in even less time. AI-3 will be smarter and more powerful than AI-2 and will quickly create AI-4, and this cycle will accelerate until, in a kinetic explosion of advancement, an uber-intelligent, nearly god-like entity will emerge.

Trans-humanists claim that on the other side of this "explosion," machines will be self-aware and humans will be able to download their memories (and personality) into one of them and obtain a form of immortality.[85]

I do not doubt that in the future some machines will be able to walk, talk, and in other ways act human; I further understand that they will be programmed to feel pain and experience some range of emotions.[86] I can even reconcile myself to equate their ability to make decisions with some rudimentary form of free will. However, there are metaphysical leaps I cannot make: I do not think AI machines will be fully self-aware or sentient, so I do not think they will be alive in a meaningful way. I do not believe that they will be capable of love, nor do I believe that I can obtain immortality by having my memories and personality downloaded into one of them.

Defining Our Terms.

–

My answer to the question, "Will AI machines be alive?" pivots on several things. One of them is the definition of "alive." It is likely that we need to move away from a binary understanding of the word. Kevin Kelly contends that AI will force us to clarify what we mean by all kinds of terms, such as *spirit, soul, consciousness,* and *mind.*[87]

Humans and machines are different. Ignore the obvious differences between flesh and blood and their gold-aluminum-silicon composition.[88] Humans delight in a child's laugh, grow melancholy on cloudy days, enjoy ice cream, and can be aroused by a lover's touch. The second may be programmed—or learn—to duplicate some of this behaviors, but that is quite different.

A machine might be able to tell you what time it is, but it cannot experience time. AI machines may appear alive but they will simply be hyper-calculating.[89]

Is Siri going to take my job? Perhaps. AI devices like Siri—a virtual assistant, or chat-bot—and more sophisticated devices are going to take over many jobs and they are going to change the ones they do not take.

In the event that comment didn't take your breath away, let me try again. A growing number of experts believe that within twenty years, many of today's jobs will be turned over to machines.

» The Obama administration believes that up to eighty percent of jobs currently paying less than $20 an hour may be replaced by AI.[90]

» Kevin Kelly argues that all jobs in which efficiency is a primary factor will be turned over to AI machines in the next few decades.[91]

» Bill Gates believes that in the next twenty years virtually all minimum wage jobs in the United States could be replaced by AI, and that few people are thinking about the serious implications this will have.[92]

» Richard and Daniel Susskind, the authors of *The Future of the Professions: How Technology Will Transform the Work of Human Experts,* believe that AI will even replace accountants, engineers, teachers, and architects, and that those not replaced will be hired, at least in part, because of their ability to interact with AI. In other words, the best doctors will be those most proficient at working alongside AI machines.[93]

I can list others who make equally unsettling claims, but I doubt I need to.[94] If you are paying attention, you are already starting to hyperventilate. Utopian novels imply that once machines do all the work we will be free to take picnic lunches and go on three-month vacations. But reality suggests a different outcome.

If an AI-bot can do my job, and do it for less money, then I'll be replaced. And if I am replaced, I lose my income. Unemployment means money problems, existential angst, and often, domestic conflict—not to mention growth of communities with large un- and underemployed workers, loss of their tax base, etc. What are we going to do?

Those who are optimistic about the future (such as Kelly) believe that accelerating technology will create as many jobs as it replaces.[95] Maybe more. Perhaps it will. But if the jobs that remain require significant education or unique skills, those looking for low-skill, good-paying jobs like they had in the 80s are in trouble.

Kelly is more optimistic than I am, but even he admits that we are headed into a season of significant disruption.

What is Work and What Is It Good For?

–

The term "work" refers to several things:

–

It addresses how we secure an income and provide for ourselves.

–

It mediates how societies' essential goods and services are produced.

–

It provides meaning to those who perform it. If machines can replace us in the first two items, the third remains.

Will the Pace of Technological Advance Continue to Accelerate?

Earlier I mentioned Ray Kurzweil's ideas about exponential change to the nth power. On the other end of the spectrum are those who argue that Moore's Law is coming to an end, and that the annual doubling of computing power will stop. The truth is likely somewhere in the middle. But short of a nuclear bomb sending survivors back to primitive conditions, it's hard to imagine technological advancements stopping. In fact, with so many technically literate young people studying science, engineering, math, and computers—and with so much money to be made with breakthrough technology—there do not seem to be many governors preventing it from charging ahead.

Even if only half of the changes I have been sharing with you come to pass, we are in for a wild ride and a disruptive few decades.

What Accelerating Technology Will Mean

Smart homes. Smarter smart phones. Virtual assistants. A refrigerator that tells the car to pick up milk. How will technology change my life? It's hard to know where to start.

The Wall Street Journal recently published *The Future of Everything,* a magazine that skipped between high-tech shoes and revolutions in golf club design to climate change and tomorrow's HR law. After looking it over, you realize that there are so many people tinkering with tomorrow that nothing will be left alone. Some of the coming changes will make our life better. Some are curiosities that we do not need and would never think about unless they show up in our neighbor's garage. Some are ominous for all manner of reasons.[96]

Here is a partial list of some of the things we can expect our future to hold.

A longer life. Not everyone thinks that our life expectancy will double again (as it has in the last 150 years), but some do. Children born today are expected to live to ninety.[97]

More ethical challenges. During the next twenty years we will add hundreds of ethical challenges to the hundreds currently unresolved. For instance, self-driving cars will be able to see an accident developing and have time to decide how to respond. In some situations, the car will need to choose who lives and who dies.[98] How will these decisions be made?

Less job security. Perhaps I've said enough about this already. But my experience is that many think their job is secure when it may not be. Consider a few examples.[99]

» In 1998, Kodak had 170,000 employees and sold 85 percent of all photo paper worldwide. Within just a few years, their business model disappeared and they went bankrupt. What happened? A technological breakthrough did away with the need for film.

» Almost overnight, Uber—who owned no cars—became the largest taxi company in the world and AirBnB—who owned no property—became the largest hotel company. People working for traditional taxis and hotels companies had very little time to realize that their jobs were being undermined.

» Today via portals to IBM Watson you can get basic legal advice within seconds, leading some to say that most legal jobs will go away.

And today, AI has the potential to disrupt both the insurance industry (as the fewer accidents of self-driving cars result in lower premiums), and automobile industry (as car makers experience a drop in traditional car sales when fleets of self-driving vehicles become ubiquitous).[100]

More terrorism. Over the last five decades, the club of nuclear nations expanded from five members (United States, Russia, United Kingdom, France, and China) to nine (the first five plus India, Pakistan, North Korea, and Israel).[101] It is remarkable that so few new nations have joined, but few feel as though we can relax. As technology expands it puts greater power of this and other types into the hands of smaller groups. It seems likely that terrorism will remain a serious problem.[102]

The expansion of "Big Brother." The age of AI will bring profound privacy challenges, especially if the threat of terrorism increases. Right now there are closed circuit TV (CCTV) cameras on many street corners—the UK having one camera for every eleven citizens—and New York beginning to follow its lead. Recently a Dallas sniper was taken out by a robotic device bearing an explosive, and some are arguing for law enforcement to make greater use of drones (including possibly arming them).[103] It's spooky to imagine anyone knowing more about our private lives than they do already, and being able to forcefully intervene more than they already can. But that is coming.

More headaches in general. Last week the WiFi unit in our home stopped working. During the three hours I spent on hold with AT&T, the three trips I made to various AT&T stores, and the time Sheri waited for a repairperson, I was reminded that the more we depend on machines, the more we depend on those who repair them. Technology requires technologically savvy support. And ironically, the more user-friendly something is, the more sophisticated the support needs to be.

Technology is great when it works. But it doesn't always work.[104] Our future will include a lot of time waiting for someone to fix things. We will need to learn patience, and so far no one has developed an app for that.

»«

I've heard that a rhino can run up to forty miles per hour but only clearly see about ten feet ahead of itself. Which means that at full speed it has a fraction of a second to react to whatever just came into focus. I doubt it's true. But if it is, the acceleration of technology will likely give us all a chance to know what it feels like to be a rhino. In our next chapter we move into the social sciences as we consider the impact of changing social, sexual, and marital dynamics.

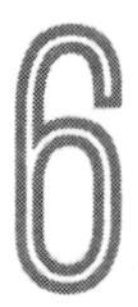

6 Changing Social, Sexual, and Marital Dynamics

One unforgettable law has been learned through all the disasters and injustices of the last thousand years: If things go well with the family, life is worth living; when the family falters, life falls apart.

– Michael Novak

[And] all the people did whatever seemed right in their own eyes.

– Judges 21:25, NLT

Just about everyone who writes about the future identifies three or four things they think will make the biggest impact on tomorrow. Most futurists agree with my decision to list accelerating technology and globalization, but they are less enthralled with my other two choices. They opt for climate change, the coming depopulation crisis, the rise (or fall) of China, debt, water shortages, and a dozen other (mostly scary) topics. I have chosen changing social dynamics and beliefs. It is to the first of these that we turn now.

My decision to focus on changing social, sexual, and marital dynamics may surprise you. It did me, although in hindsight I realize that it shouldn't have. It's actually one of the main reasons I was prompted to start looking ahead.

A Fateful Day

As I mentioned earlier, I serve as the pastor of a local church. One day about two years ago I stepped out of my study to find out what some of the other staff were up to. My assistant said, "David is talking with some parents who are worried about their high school daughter. Jeanne raced out to counsel someone who is suicidal, and Syler is talking with a single mom who just lost her job."

None of that sounded unusual. Part of being a pastor is helping people in crisis. It's what she said next that got my attention. It turns out that none of the people they were helping attended the church.

This was consistent with a trend I'd been noticing. More people we did not know were knocking on our door when their lives were unraveling. Since we're glad when someone takes a step back towards God—and because we know that a crisis is one of the most common reasons people do so—I was encouraged to hear that "we are busy." But one of my first thoughts that day was, *How are we going to make this work? If the number of people looking for help keeps growing, we are not equipped to handle what is going to be walking in our door.*

Not long ago I met with a pastor in his late seventies. Over coffee he commented that when he started in ministry fifty years earlier, it felt like about eighty percent of the families in the area were reasonably healthy, and less than twenty percent were struggling. "But over the last sixty years the ratios have switched," he said. "And the twenty percent that appear to be making it are running hard and fast trying to survive themselves."

I labeled this glacier, "Changing Social, Sexual, and Marital Dynamics." If I were willing to be more alarmist, I would call it something like, "The Coming Collapse" or "The Breakdown of the Family."[105] The challenges on this front actually cause me more angst than cyber security issues or climate change.

In an effort to keep this simple, I am going to limit this chapter to five points, focus exclusively on how this issue is playing out in North America[106] and remind you that the endnotes allow you to track this argument in greater detail.

Let me also start with an apology. Discussions that include sex, marriage, divorce, living together, parenting, illegitimate children, broken homes, and related terms can become heated. They can also be hurtful.[107] It is not my intention to offend anyone. I am simply trying to peer around the corner. And I can offer four high-level observations.

One: Marriage and Family Are Mission Critical

Marriages and families matter. A lot. When they are strong, people can face almost any challenge. But when they are weak, it doesn't take much to pull people down.

At various times in the past, people have tried to do away with both of them. It has never worked. These efforts have not only failed. They have failed quickly.[108] Marriage is too foundational to be pushed aside. And the family not only operates as the center of life and source of nurture, it also serves as the dispenser of culture, haven in a storm, and several other things besides. As one pundit quipped, "The family is the original Department of Health, Education, and Welfare."[109]

Though it may seem uncaring to say this, children do best in stable, two-parent homes. Of course having two parents is no guarantee of success, but it improves the odds. Children who grow up without a father are five times more likely to live in poverty, five times more likely to commit a crime, nine times more likely to drop out of school, and twenty times more likely to spend time in prison.[110]

And it's not just individuals who do better when the family is strong; governments do as well. When families are healthy, the state operates from a foundation of strength. But when families unravel, the state is left with fewer citizens, greater poverty, increased crime, and higher welfare spending.[111]

Two: Recent Changes Are Unprecedented

Though some think "the family" has always been Mom, Dad, 2.4 kids and a dog named Spot, that is not true. Marriages and families have always adapted to the times. Centuries ago, marriages were arranged, polygamy was an option, and extended families were the norm. Today matchmaking is computerized, polygamy is illegal, and families have been downsized. The family has adapted as needed to fit into the challenges of the moment.[112]

But the changes that have taken place over the last fifty years are different. On a scale of 1 to 5 they come in around 8.2. Moral values have relaxed, marriage has been repositioned, and the family has changed so substantially that it is now difficult to define. Beginning with "The Pill" and continuing through the advent of gay and lesbian marriages, there has been a revolution in social, sexual, and marital dynamics.[113]

Compare and contrast. You see these changes when you compare marriages from one hundred years ago with today. Back in the early twentieth century, marriage was understood as a sacred covenant and legal contract entered into between a man and woman who were committed to living and working together for the rest of their lives. It was assumed that he would provide for her; she would run the home, and together they would raise children and care for each other until one of them died.

At that time: marriage assumed both monogamy and sexual fidelity; there were well-established gender roles; getting married was something that happened to nearly everyone; women were expected to marry when they were in their late teens or early twenties; both males and females were expected to be virgins on their wedding night; and as soon as they married, a couple would begin having children.[114]

It was also understood that one of the reasons you got married was to help everyone else. The unstated assumption was: marriage leads to sex; sex leads to babies; and society needs babies, therefore society needs marriage.[115]

If you advocate that model today, be prepared for a visit from HR. It's not that the "traditional view" is no longer out there. It is. But the assumptions surrounding marriage today are much different. As opposed to the earlier view, what you are likely to hear now is that marriage is a voluntary arrangement made by two people who love each other and who believe that being together offers them their best chance for personal happiness and fulfillment.

Today people believe that:[115]

» The goal of marriage is the happiness of the couple, not the raising of children.

» Marriage is a private agreement governed by the couple's desires, not by the needs of society or the dictates of government.

» Marriage may be a gateway to adulthood, but it is not the only one, and it is neither expected nor required.

» Those who marry before their mid-twenties are foolish and their marriage will almost certainly fail. Both spouses need to finish their education and get established before they settle down.

» Entering marriage as virgins is unrealistic. Besides, living together is a wise first step because it allows a couple to see if they are compatible.

This leads us to the third main point.

Three: The Sexual Revolution Changed Just about Everything

There are a number of reasons why marriages are different today. For starters, it is no longer necessary in the way it was earlier. We take surviving for granted. Our grandparents were not so fortunate. Their grandparents were even less so. Through the early part of the twentieth century, life was hard and people married to survive. It was less about romance and more about reality. Most people were farmers and farming required physical labor. On top of this, running a home was challenging. The solution was simple: men did the heavy lifting, women attended to the cooking, cleaning, and childcare, and by working together they built a life.[116]

Sexual practices have also changed. For starters, marriage is no longer necessary for sex. In the past, many people got married to have sex, or they got married because they'd had sex and the woman was pregnant. Of course there were exceptions. People have always had sex outside of marriage. But massive changes happened during the '60s in both practices and attitudes. Today marriage is no longer viewed as the only legitimate channel for sexual expression. In fact, sex before marriage is the norm.[117]

A further reason marriage is so different today: children are no longer an asset. One hundred years ago they were. Young children served as unpaid farmhands and older children functioned as 401k plans. Consequently, a married couple wanted as many children as possible.

Today young children are an economic liability and the government has promised to take care of us in our old age. As a result, most couples now have one or two children. This changes just about everything: women no longer spend their adult lives pregnant and caring for small children; women are much less likely to die in childbirth; both the husband and wife are likely to live longer (which puts new strains on marriages); because women are likely to have a full life in addition to being a mom, many are likely to pursue more education than in the past, which means they will be older when they marry.[118]

I could list a half dozen other reasons why marriage is different today. But in order to understand what is likely to happen in the future we need to explore the factor that has changed marriage more dramatically than any other—the "sexual revolution."[119]

The impact of the sexual revolution. Most people are aware that the sexual revolution was a big deal, but it's easy to underestimate its impact.[120] Did you know that:

» ***Before the revolution, Louis Réard could not find a bikini model.*** In 1946, Reed chose to name his new swimsuit "the bikini" after an island in the South Pacific where the United States had tested atomic explosions. He did this because he wanted to "set off a bomb" in society. Unfortunately for Reed, he couldn't find a single woman willing to model the suit. Even the French thought it was too risqué. Reed eventually solved his problem by hiring a stripper. Today few of us can imagine the controversy. Indeed, this year eight billion dollars' worth of bikinis will be sold.

» ***Before the revolution, TV married couples slept in separate beds.*** In 1961, Dick Van Dyke and Mary Tyler Moore played Rob and Laura Petrie in a popular, evening prime time TV sitcom. Even though they were married in the series, they were shown sleeping in separate beds because studio executives thought showing them in the same bed would be scandalous.

» ***Before the revolution, "Heff" was repeatedly written up for obscenity.*** In 1963, ten years after Hugh Heffner published the first edition of Playboy magazine, he was still battling the courts over charges that the photographs of topless women he was publishing were obscene. In the early '60s he even briefly went to jail.[121]

The sixties did not simply accelerate an ongoing liberalization of sexual norms, it radically altered them.[122] Today's bikinis are skimpier than Reed could have imagined; the only people not sleeping together on TV are people who are married to each other; and Heffner's foray into pornography has turned into a movement that now rivals professional sports in terms of revenue.[123]

Let me be clear, the sixties did not invent illicit sex. But it was during this time that sexual practices and societal values became untethered from the past. And in the process, social, sexual, and marital dynamics all changed.

So what does the future hold? In order to answer this question, we need to move to the fourth point.

Four: We're Now Experiencing Second- and Third-Order Effects

My high school health class had a unit on "Sex Ed" that included a guest lecture on "Safe Sex." Though that was over thirty years ago, I was a fifteen-year-old boy and the topic was sex, so I paid rapt attention. I can easily summarize what the speaker said.

> "I am required to advise you not to have sex until you are

Sex and the University.

–

One of the places where it is especially easy to see how our sexual values have changed is in higher education. Consider Yale.

During the first 150 years of her existence (1701-1850), Yale was an all-male school that focused almost entirely on religious training. The founder of the institution, a famous pastor named Cotton Mather, even described the school's principle mission as ensuring that every student strives to "know God in Jesus Christ and answerably lead a godly, sober life."

During the century between the Civil War and the 1960s, most colleges became co-educational but the sexes were kept far apart. Yale was no different. Men were never allowed into the women's rooms and the hours during which women could visit men were restricted. On top of this, dates were chaperoned and Yale assured the student's parents that the school would act in loco parentis (i.e., it would police student life, protect the sanctity of their daughters, and cultivate moral character in all). »

---

married because it could lead to pregnancy, a sexually transmitted disease or both. But I want to say something else. All of the downsides can be mitigated if you use a condom. No pregnancy. No STDs. As long as you use a condom, nothing bad can happen."

Over the last fifty years, colleges have distanced themselves from their earlier morality statements and the barriers between the sexes have been removed. When women were first admitted they had their own dorms. Over time, dorms became co-ed by floor. Then they became co-ed by room—i.e., girls would live in a room next to a room full of boys. Today boys and girls share the same room, and the university not only does not object, it often ostracizes those who do.

Beyond this, there is Sex Week, a week-long campus festival which not only includes film screenings sponsored by the sex industry, but also features lectures and demonstrations by porn stars.[124]

It's unlikely that Cotton Mather envisioned Sex Week when he founded Yale. And it is impossible to explain the transition of the school without pointing to the sexual revolution.

Ten years later, as a young college pastor serving on a campus that officially sponsored "Outdoor Intercourse Day," I saw things differently. Lots of people were having sex, and many expressed no reservations. But there was a dark side, especially among the women. It turned out for them that casual sex was not so casual.

I grew so frustrated by the pain this was causing many students that I eventually started speaking out on the topic. My claim was that sex was designed to lead to a profoundly intimate, vulnerable, emotional, and spiritual bond that would unite two people body and soul. I also said that in light of this, sex was best expressed in a relationship with someone who is committed to your long-term best interest (i.e., a spouse).

The argument was never very popular. But I still think it's valid. In fact, twenty years later I am more convinced than before, because as a pastor I now deal with those who slept their way through their twenties (and sometimes their thirties) and are hampered by their past.

But my biggest take-away is not over what I got right thirty years ago, but over what I missed. Though I saw the toll casual sex was having on people, I did not see the toll it was having on society.[125] Let me explain:

"The Pill" allowed us to bypass previous consequences of sexual freedom. In the past, if a couple slept together, she often became pregnant. And though sexual activity can be kept secret, babies cannot. As a result, natural consequences kept a lot of sexual activity in check or led those who were sexually active to get married. The development of oral contraceptives changed this. As a result, a dramatically new sexual freedom emerged.[126]

We are now facing a further set of effects. If normal cause and effect was in place (e.g., sex leads to pregnancy and marriage), at least some who are currently sexually active would choose not to be, and many of the couples who were active would eventually be "expecting" and no longer be "available"—i.e., the sexual marketplace would be different).

Because these effects were removed, sexual freedom continued unchecked for several decades. My argument is that we are now experiencing second- and third-order effects. What am I talking about?

Effect one: Fewer people are getting married. In 1960, more than 70 percent of all adults were married, including nearly sixty percent of twentysomethings. Today, just twenty percent of those between eighteen and twenty-nine years old are married.[127]

Why the drop-off? There are a variety of reasons, but let's not miss the obvious one: when commitment-free sex is available, many choose that over marriage.[128]

Most people still want to get married, but as Figure 6.1 shows, the number of people who marry has steadily declined over the last few decades.

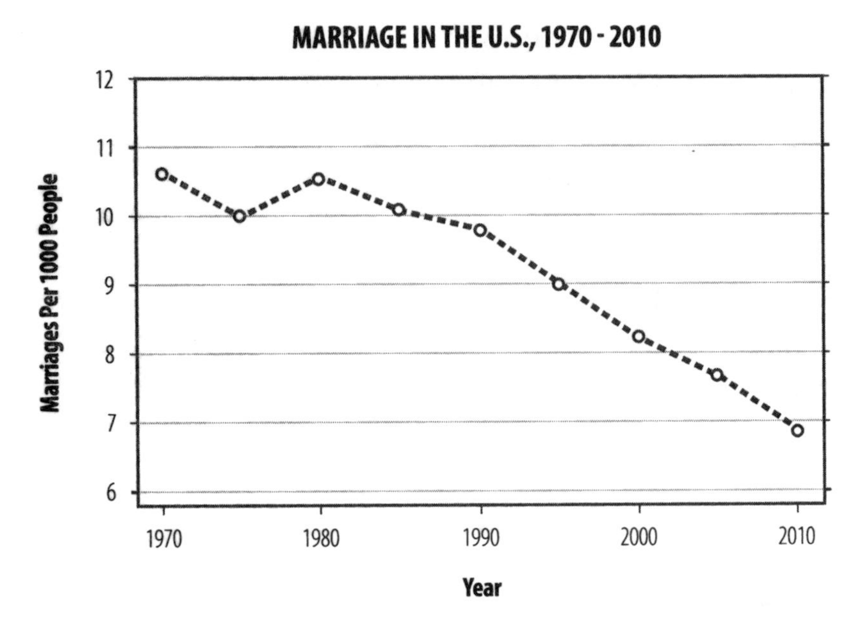

Figure 6.1 The drop in frequency of marriage in U.S. over the last forty years, a drop by approximately one-third of the 1970 level (2015 analysis by R.S. Olson, CDC NCHS data).[129]

We are now in uncharted waters. For the first time in the history of the United States there are more single adults than married ones.[130]

Marriage used to be the norm. Now it is the exception.

Effect two: The birth rate has dropped. People are not just avoiding the altar; they are also avoiding the maternity ward. Given that both United States and global populations continue to climb, I was unaware of this "death of birth." But the number of children being born has dramatically dropped during the last fifty years, and has continued to decline over the last twenty, to a current record low (Figure 6.2). In fact, in almost every Western nation it is now well below replacement rates. The only reason the global population is climbing is because everyone is living longer.

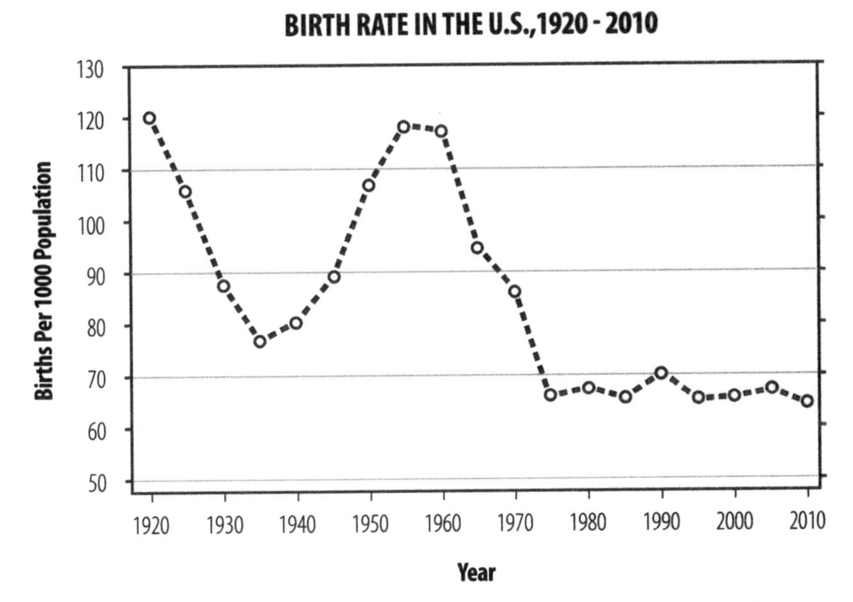

Figure 6.2 The drop in the U.S. birth rate over the last ninety years, a drop of approximately 40% of the 1960 level (2012 Pew analysis, data of the CDC NCHS and R.L. Heuser).[131]

The American family has never been smaller.[132] This will have significant long-term social, emotional, and economic implications.

Effect three: Children born today are increasingly born outside of marriage. Not only are fewer children being born, but a growing percentage of those who are, are being born to single moms. As Figure 6.3 shows:

» In the 1930s, about five percent of children in the United States were born outside of wedlock.

» In 1965, when White House Counselor on Domestic Policy, Daniel Patrick Moynihan, sounded his alarm about the rising number of out-of-wedlock births, this national statistic was still below 10 percent.

» Today 40 percent of all children—and 50 percent of Hispanic, and 70 percent of African American infants—are born to single moms. In some urban settings the number is approaching 80 percent.[133]

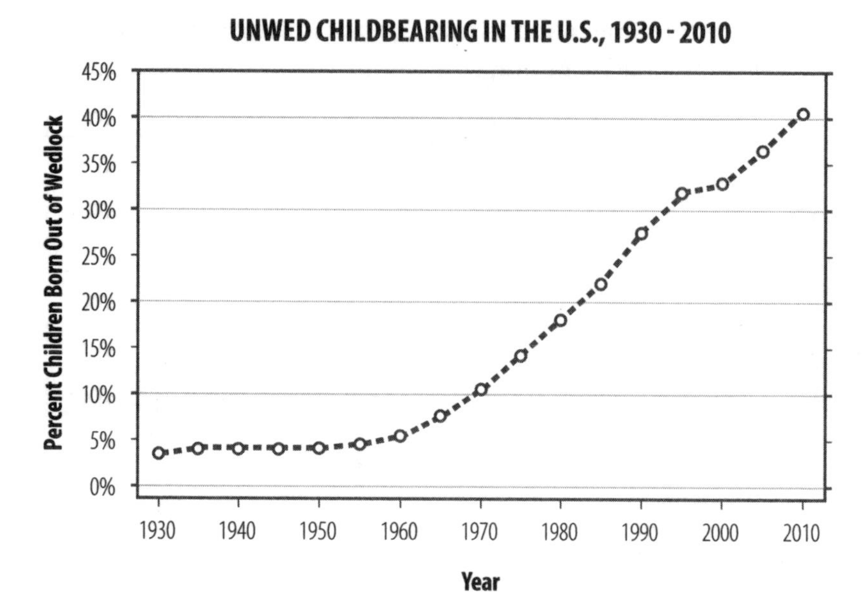

Figure 6.3 The rise in out of wedlock births over the last eighty years, an approximate 8-fold increase over the 1960 level (NYT analysis, CDC NCHS National Vital Statistics data).[134]

I stand in awe of the herculean efforts being made by single moms (and dads) everywhere. They deserve our prayers and support. But I want to note that children who only enjoy the daily support of one parent often struggle. University of California-Riverside researcher David H. Bailey, a Senior Scientist emeritus at Lawrence Berkeley National Lab and popular math and science commentator, summarizes the situation when he writes:

> A litany of statistics underscores the difficulties of both single parents and their children. Children born to unmarried women are more likely to grow up in poverty (particularly in the United States), have lower-than-average educational attainment and become unwed parents themselves.... Forty-seven percent of unwed mothers in the United States are at or below the poverty level; by comparison, even mothers who marry after conception experience only a 20% poverty rate (2002 data). The percentage of households unable to meet basic expenses is 30-36% for cohabiting and single parents, but only 15% for married couples...[135]

So where does this leave us? I do not see much good coming our way. Charles Murray, a political scientist, shares my view and explains his concerns in *Coming Apart: The State of White America, 1960-2010,* where he argues that the twenty-first century American middle class is dividing in two. There are those in the "New Upper Class"—who pursue a college education, marry, and then have children—and who tend to stay married and enjoy both a stable family life and a decent income. Then there are those in the "New Lower Class"—who do not graduate from college, and do not wait until they are married before having children (or simply do not marry)—and who often struggle. Lacking the structure marriage brings, their lives are generally more disordered, and their children are "often passed around to various persons who are moving in and out of a stable family life"; unsurprisingly, "the children of such families have sharply diminished life chances."[136,137]

Murray develops his thesis over several hundred pages, illustrating it with lots of charts and graphs. I am not going to repeat any of it here, other than these two points. First, the division between the Haves and Have-Nots is occurring as a result of different practices surrounding family, faith, and work. Those who have embraced traditional family forms are doing much better than those who have not. Second, those who are doing well embrace traditional values even when they advocate more progressive ones.[138,139]

As with any social trend, there is a debate over what is really happening and why. At this moment, those on the left argue that poverty is forcing people to make poor choices while those on the right contend that poor life choices are leading people into poverty. I'm not going to referee that fight here. I simply want to note that this path has been traveled before. History suggests that cultures start with a conservative sexual ethic, relax it over time, and then face challenges as things unwind.[140]

Dale Kuehne, a political scientist at Saint Anselm College, notes that cultures that collapse tend to have two things in common—relaxed sexual values and escalating debt. In fact, Kuehne not only argues that cultures in decline all have debt and liberal sexual practices, he makes a stronger claim: when debt escalates and sexual norms relax, societies will crash.[141]

We need to consider all of this as we project the trend line. But before we go there, I want to pause briefly for a reality check.

A Reality Check

Because discussions concerning sexual ethics tend to happen in echo-chambers—on one side are those who describe today using terms like "depraved" and "sewer." And on the other are those who suggest that anything that hints of tradition is "oppressive" and "misogynist"—I'd like to suggest we celebrate the things that are better while taking a clear-minded look at where we are headed.

To those who pine for the good old days, let me remind you that they were not as good as you think. Or, to quote Will Rogers, "Things aren't what they used to be and probably never were." Matt Ridley, an Oxford professor, member of the House of Lords, and a popular journalist, sets this up by comparing our perception of the past with reality. The situation he describes is not focused on sexual ethics, but he makes a profound point all the same, and I think the overlap is clear.

Ridley starts by asking us to imagine a better-off-than-average family living somewhere in Western Europe in 1800.

> The family is gathering around the hearth in the simple timber-framed house. Father reads aloud from the Bible while mother prepares to dish out a stew of beef and onions. The baby boy is being comforted by one of his sisters and the eldest lad is pouring water from a pitcher into the earthenware mugs on the table. His elder sister is feeding the horse in the stable. Outside there is no noise of traffic, there are no drug dealers, and neither dioxins nor radioactive fallout have been found in the cow's milk. All is tranquil; a bird sings outside the window.

He then fills in the details, which seem to be edited from our "Currier and Ives" mindset. Ridley writes:

> Oh please! Though this is one of the better-off families in the village, father's Scripture reading is interrupted by a

> bronchitic cough that presages the pneumonia that will kill him at 53—not helped by the wood smoke of the fire. (He is lucky: life expectancy even in England was less than 40 in 1800.) The baby will die of the smallpox that is now causing him to cry; his sister will soon be the chattel of a drunken husband. The water the son is pouring tastes of the cows that drink from the brook. Toothache tortures the mother. The neighbor's lodger is getting the other girl pregnant in the hayshed even now and her child will be sent to an orphanage. The stew is grey and gristly yet meat is a rare change from gruel; there is no fruit or salad at this season. It is eaten with a wooden spoon from a wooden bowl. Candles cost too much, so firelight is all there is to see by. Nobody in the family has ever seen a play, painted a picture, or heard a piano. School is a few years of dull Latin ... Father visited the city once, but the travel cost him a week's wages and the others have never travelled more than fifteen miles from home. Each daughter owns two wool dresses, two linen shirts and one pair of shoes. Father's jacket cost him a month's wages but is now infested with lice. The children sleep two to a bed on straw mattresses on the floor. As for the bird outside the window, tomorrow it will be trapped and eaten by the boy.[142]

Things are better on so many fronts we would be foolish to think otherwise. However, we are equally foolish to deny the harm caused by the sexual revolution. Does anyone think it's a good idea that little girls dress like sirens or that there are apps to facilitate random sexual hook-ups? Do Disney characters really need to sport cleavage? Can no perversity go un-videoed; and can no perverse video go unaired?

Children fifty years ago did not grow up sexting or watching erectile dysfunction ads. Do we want them doing so today? Shouldn't we try to imagine where today's sexual ethics might lead and chart a saner course? There has to be a category between profligate and prude.

Where Are We Headed?

Will this trend continue? I believe it will.[143] In fact, I agree with professor Jeremy Neill, who said, "If I were a broker and the sexual revolution were a stock, I would still be urging my customers to buy shares."[144] I expect the sexual revolution will eventually end because history suggests that libertine experiments implode.[145] But I do not think the end is imminent. Between now and then I believe several things are likely.

The definition of marriage will expand, but the number of people getting married will decline. The legal arguments in the same-sex marriage decision, Obergefell v Hodges, will likely be applied to group marriages, temporary marriages, and other arrangements.[146] However, though this will allow more people to marry, I believe fewer will.[147]

The number of children will continue to decline. Though the United States will fare better than any developed country concerning population decline, we are currently below replacement rate and I do not foresee a baby boom in our future.[148]

The declining number of children will cause economic hardship. Fewer people mean fewer workers, which means a smaller GDP, a smaller military, less money to invest in infrastructure, and eventually declining real estate values. In an effort to keep the tax base stable, governments will take certain steps (i.e., offer incentives to women to have children, expand state-run day-care, expand maternity and paternity leave, roll out the red carpet to immigrants, etc.), but these will have modest impact.[149]

The middle class will continue dividing. When things become challenging, those in stable families are best able to weather the storm. I believe the New Upper Class will continue to fare better than the New Lower Class.[150]

Government will grow. There are three institutions that provide stability: the family, the church (synagogue, mosque, etc.), and the state. In the West, the first two are losing influence, which means the third must grow. As Figure 6.4 below suggests, the United States government has been growing; indeed, government spending is currently higher than at any point, including World War II.[151]

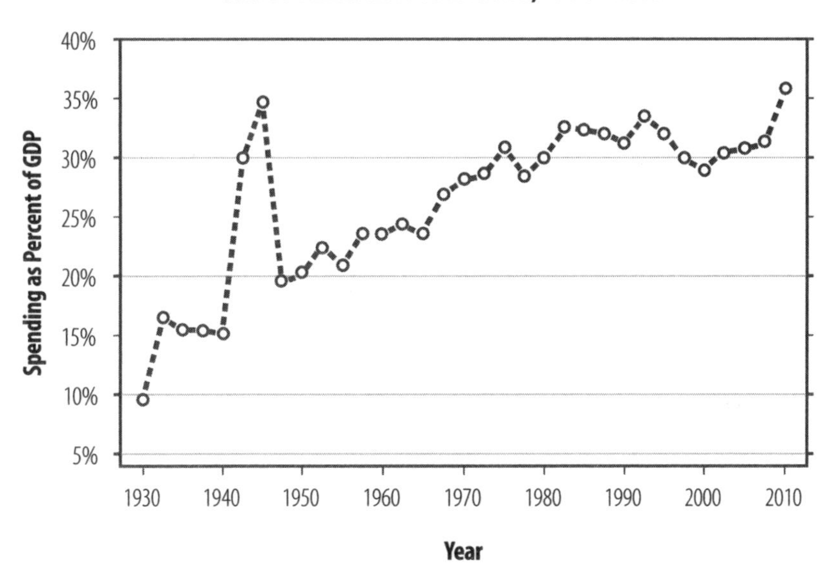

Figure 6.4 The rise in government spending, GDP adjusted, an approximate 3.5-fold increase between 1930 and 2010 (2015 analysis by R.S. Olson, of Bureau of Economic Analysis data).[152] Note the sharp peak corresponding to 1945, a high that has recently been surpassed.

The U.S. federal debt will increase. Of course, when the government grows, federal spending increases and debt follows. This is the current pattern. In 1970 the United States federal debt was zero. As of July 2016, it was 19 trillion, and by some estimates our total indebtedness (including Social Security and other "unfunded liabilities") is 70-90 trillion.[153]

You do not need a degree in economics to spot the trouble: 1) the United States has an aging population; 2) retirees pay less in income taxes but make greater demands on the system; 3) an aging population is especially problematic for governments with existing debt and unfunded pension plans; and 4) efforts to bridge the gap will force the government to raise taxes on the young, cut benefits for the elderly, and reduce capital investments and services (schools, bridges, police, and parks).

I could go on, but you get the point. We face challenges in the years ahead.[154] Which leads to question: what should I do? Is there a way forward? Yes, I think there are a couple helpful things you can do besides the obvious things that emerge from this chapter (i.e., stay out of debt, avoid the long arm of the sexual revolution, and adjust your expectations for the future).

I encourage people to intentionally invest in their family and friends. Social problems benefit from social solutions. A strong marriage, a strong family, and/or good friends may be the best way to position yourself for the years ahead, and allow you to help care for those who are not fairing as well.

»«

In the next chapter we will consider the effect that globalization is having in the world.

Globalization

This thing called globalization can explain more things in more ways than anything else.

– Thomas Friedman

Globalization is not a monolithic force but an evolving set of consequences—some good, some bad, and some unintended. It is the new reality.

– John B. Larson

A few months ago, as I was sitting in an airport lounge in Chennai, India, on my way to Nepal, I saw globalization on full display right in front of me. Immediately to my right was a Krispy Kreme donut stand. To my left were advertisements for Earl Grey tea, Hyundai autos, and Rolex watches. And right in front of me was an Indian boy wearing a Ronaldo jersey.

In other words, while flying to Nepal I was surrounded by American donuts, and ads for British tea, Korean cars, Swiss watches, and an Indian boy wearing the jersey of a Portuguese soccer star who plays for a Spanish team.

Welcome to the Global Village. Today our world is interconnected in political, economic, cultural, and environmental ways previously unheard of, and these links are changing everything. The term for this phenomenon is globalization.

Thomas Friedman, a best-selling author and the Pulitzer Prize winning columnist for the *New York Times,* defines the phenomenon this way:

> Globalization is the combination of technological advances in communication and travel, coupled with the aggressive spread of capitalism and democracy, that is making borders mean less than they used to.

He goes on to argue that beginning around the turn of the century, globalization replaced the Cold War as the overarching framework shaping our world.[155]

Objection Overruled

In the event you think the pundits are overstating globalization's impact, it's worth looking around yourself.

» Unless you live one hundred miles from the nearest gas station, there is a Mexican restaurant close to your home. Maybe two. And a Japanese one as well. And Thai. And Brazilian. And some new concept restaurant featuring a fusion of three cultures you've never heard of.

» It's likely that your shoes were made in Malaysia. Your smart phone has components manufactured in China, Korea, Australia, Germany, Italy, Spain, and Texas.

» The person you dueled in online chess last night lives in Bolivia, and the person who helped reconcile your Visa bill this morning lives in India. And you know more about what is going on in the Middle East than you do with your neighbor who lives across the street.

It wasn't like this forty years ago. We did not have global news channels or free long-distance calling plans. Most of our food was grown within one hundred miles of our home and people lived their whole lives without ever leaving their country. For that matter—and this is a big matter—our economy was not as internationally integrated as it is now. Things have changed.[156]

"Toto, I'm pretty sure we're still in Kansas, but it looks like L.A., Hanoi, and Hong Kong, with a bit of Rio mixed in."

Understanding Globalization's Impact

Globalization is not a new idea. The Phoenicians practiced international commerce almost as early as civilization took root. In the late thirteenth century, Marco Polo left Venice to open a trade route with China. And "in fourteen hundred ninety-two, Columbus sailed the ocean blue," looking for a shorter route to the spice-rich East Indies. International trade has had a long on-ramp.[157] It's not the newness of globalization that is significant. It is the scope, scale, and complexity of today's international interaction that is unprecedented. Technology, transportation, and trade have shrunk the planet. The tall walls that used to separate countries and cultures are not as significant as they once were.[158]

Some think this is a bad thing. They argue that "globalizers" are overrunning native cultures, exploiting the poor, fueling terrorism, degrading the environment, and elevating human insecurity.

Others think globalization is a good thing. They contend that it creates jobs for the poor, and contributes to the development of the infrastructure of less industrialized nations.[159]

East Falls to West.

–

Though there are still a few wire-haired academics singing the praises of central planning, anyone paying attention knows that Adam Smith bested Karl Marx. The evidence is everywhere. In the mid-1980s there were 45 democracies in the world. Today there are 120 and they represent the most vital economies on the globe. The Soviet Union collapsed, the wall came down. And rather than clip Hong Kong's wings, China has allowed her to fly economically, and attempted to follow behind. One system works and the other does not. To put a finer point on it: one system gives you Google, the other leads to North Korea. The eight-track tape has a better chance of making a comeback than communism does.[160]

Both are right. Globalization is a complex mix of good and bad, and your view of it depends upon who you are, where you live, and what you do. My goal is not to advocate for or against it. Right now I'm simply trying to figure out where it is taking us. There are six things to be aware of.

First: Globalization largely leads to westernization. Those parts of the world operating under radical Islam continue to stiff-arm the West. Furthermore, the West is now more global (and thus a bit less "west") than it was twenty years ago. However, the battle between East and West is over and the West won, America in particular. English is the lingua franca of the world. Representative democracy is the go-to political system. Kids everywhere eat at McDonalds, wear Levis, and drink Coca Cola.

Think about it for a moment. Ronald McDonald accomplished what neither Napoleon nor Hitler was able to do—he marched into Moscow and set up shop.[161] It's not entirely fair to say that globalization equals Westernization, but it's not far off.

Second: Globalization spreads capitalism. Globalization and capitalism are joined at the hip. The first doesn't work without the second, and when you combine the second with today's communication and transportation technology, you end up with the first.[162] There is much about this situation to celebrate. For starters, the spread of capitalism has led to a significant reduction in the number of people living in extreme poverty.[163]

However, there is a dark side as well. Winston Churchill

famously complained that "democracy is the worst form of government except all those other forms that have been tried." Some say the same thing about capitalism. My assessment is that a free market shaped by compassion works well—think George Bailey from *It's a Wonderful Life.* But capitalism absent compassion is something quite different. It is an efficiently ugly machine that rolls over people for profit—think Mr. Potter.[164]

Third: Globalization fuels trade and transnational corporations. Organizations that learn how to move fast, wow customers and gain market share are now able to generate wealth in one hundred countries instead of one or two. And those that dominate their industry across international markets are able to grow larger than some countries. Wal-Mart is now bigger than Norway, Visa is bigger than Zimbabwe, Royal Dutch Shell is larger than Egypt, and Apple is gaining on Saudi Arabia.[165]

Fourth: Globalization creates instability. Capitalism provides more products and more choices for less money, however it does so by eliminating the slow and weak. The rough justice of the market means that there will be losers as well as winners. Capitalism does not coddle.[166]

In *The World Is Flat,* Thomas Friedman argues that whatever tension may exist between nation-states, the ongoing war is now being waged in the marketplace. And it's not just corporations that have donned fatigues and started sleeping in foxholes. It's individuals as well.

Friedman claims that if you were given a choice between hiring a B student in Boston or an A student in Bangalore, thirty years ago you went with Boston. But not today. With the technology now available, anyone with a laptop can contribute from anywhere in the world.

> When I was growing up, my parents told me, "Finish your dinner. People in China and India are starving.' I now tell my daughters, 'Finish your homework. People in India and China are starving for your job."[167]

As globalization expands, competition increases. Companies—and their employees—not only have to stay ahead of their competition across town, they now have to run faster

We Promote Globalization Whether We Intend to or Not.

–

It's not uncommon for people to speak out against globalization (and capitalism). Perhaps you have. I understand. The massive scope of these systems and trends has downsides and it creates hardships. But just about everyone is fueling the process, including the critics. How? By wanting the best deal for their money.

We want the highest quality product we can buy for the least amount of money. And we want it when we want it. In blue. No, change that to green.

It's not that we yell if we do not get our way. We simply shop somewhere else. And even then we remain fickle. We are always looking for the best deal. »

and jump higher than the start-ups in Vietnam, Sri Lanka, and a dorm room in Austin, Texas.

This creates a win for customers. After all, if they want high quality, low-cost products global-capitalism is their best friend. Greater competition means lower prices and better results for customers.

But it means havoc for companies and employees. The same person who loves buying cheap sunglasses is devastated when the factory she works at relocates to Manila.[168] The product will now be available to customers for less money and the company will realize a bigger profit, but the job has left the neighborhood. And once it leaves, it is not coming back. (Although as soon as wages rise in Manila the plant will be moved to Hanoi.)[169]

Globalization accelerates the process of creative destruction, and in doing so it destabilizes lives.

Five: Globalization fosters complexity. Years ago you could monitor your local economy by paying attention to what was going on around you. If you knew the cobbler and were impressed by his hard work and honesty, but noticed that his kids were wearing ratty clothes and looked a little thin, you bought another pair of shoes and paid full price. You wanted the cobbler to do well—indeed, you needed him to stay in business, and so you made ongoing adjustments to the system to make it work.

That doesn't happen today. We may say we live in a global village, but it's more global than village. I have no idea who made the shoes I just bought or what they were paid to do so. I temporarily boycott any corporation I hear is operating sweatshops in some faraway land. But the system is so massive I have no idea how it works and no ability to tweak it.

Six: Globalization elevates consumerism over citizenship. John F. Kennedy famously encouraged Americans to "ask not what your country can do for you—ask what you can do for your country." George W. Bush asked that we support the Gulf War by going shopping. Somewhere between presidents, patriotism lost ground to the market.[171]

This process fuels globalization whether we intend it to or not. Our purchasing habits force organizations to experiment and innovate to keep our business. And they force old, slow, and inefficient businesses to close down. The end result is that products get better (e.g., cars replace horse-drawn carriages, cell phone batteries last longer). Our collective standard of living goes up. But there are losers. Blacksmiths and buggy-whip makers go out of business. Real people get hurt.[170]

There are more objective ways to document the growing influence of the market (as opposed not only to patriotism but also the family). One hundred years ago a person saw a few low-key ads every month. In the seventies the number had spiked to five hundred per day. Today it's five thousand.

Think about that. Five thousand ads per day. Every day. If you had been born one hundred years ago, you would see a few thousand, low-key informational ads over the course of your life. Because you are alive at this moment and in this culture you will be exposed to tens of millions. And few would describe what we see today as "low-key" or "informational." Sophisticated, compelling, and subliminal maybe, but low-key and informational, probably not.

A bunch of smart people woke up this morning motivated to get you to buy things you do not need. And now that technology helps them track your buying habits—and access your Google searches—they will be able to design even more effective appeals tomorrow.[172]

The Globalization-Capitalism dance is changing us from citizens of a country to shoppers in a mall. Logos are replacing flags.

Will Globalization Continue?

Because the world is still significantly Balkanized, trade and travel can expand. I believe it will. The recent vote by Great Britain to leave the European Union suggests that movement in this direction will face significant headwinds, but globalization has stalled before and rebounded. Absent a major catastrophe or war, the trend line of history suggests ongoing global integration.[173-175]

What will this mean to us? I am expecting four things:

More tension. If everyone in your neighborhood moved into a single home, you would need new rules to govern how loudly people can play music, how often the trash needs to be emptied, and what happens to those who do not clean up the messes they make. With these policies you might avoid killing each other, but tempers are still apt to flair. Globalization is this neighborhood writ large. The tall walls separating countries and cultures are coming down and we need to find ways to get along. Harvard political scientist Samuel Huntington does not think we can do it. He predicts a "clash of civilizations" between Islam and the West. Former British Prime Minister Tony Blair and *New York Times* columnist Thomas Friedman are more optimistic. The first believes we will find a way because we have to. The second thinks we're unlikely to go to battle if it will lead to disappointing earning statements for large corporations.[176]

Bigger winners. Globalization allows winners to win on a bigger stage. Think Starbucks. In less than fifty years the coffee chain has grown from a single Seattle outlet to twenty thousand stores in sixty countries. As I've already noted, some love this, others decry it for homogenizing cultures, crushing locally-owned establishments, and spreading Western values. Whichever camp you land in, Starbucks is a textbook example of winning big. And there are may other besides.[177]

Frustrated others and more hurry. Of course, where there are winners there are losers. Globalization leads to economic marginalization for many, which elevates tension levels even more. Globalization also means you will likely be more rushed tomorrow than you are today, both because you have more options and you have more competition.[178]

More ads. The partnership between technology, retail, and big data means more people will be trying to sell you more things more often. And they will get better at it all the time, because they'll know what you like, what you bought last, and when you will likely need to buy again. In fact, they may design an algorithm just for you. Don't worry too much about Big Brother. Google, Amazon, and the local grocer are the ones watching you most closely.[179]

How will all of this affect you? In addition to the things just mentioned (more tension, competition, and distraction), and a few other obvious matters (more travel, access to diverse foods, a flood of low-cost goods), *I think globalization will start to reward those who cultivate quiet.*

We will not be able to run as fast as globalization or accelerating technology demand. And to try to do so will not work. I am advocating almost the opposite—instead of learning to move faster, we must work on cultivating an inner depth and calm. I've already told you that I serve as a pastor, so it's no secret where I think you look for this. (Hint: look up, not in.) I am not sure what Jim Collins's advice is as we move into a world that is more "ferocious."[180] But my day starts with prayer, Scripture reading, perhaps some journaling. I believe the way for each individual to navigate globalization is to cultivate an inner world that is strong enough to shape our outer one.

»«

In the next chapter we will turn to the final glacier, which deals with the current and future impact of swirling ideologies and religious beliefs. There I will be using the term "faith" to refer to these beliefs—in a broad and generic sense, to include organized religions, but also various other personal ideologies, animating ideals, and ultimate convictions.

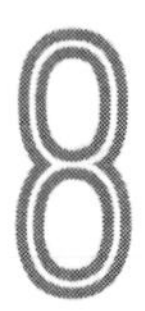

8 The World of Swirling Ideologies and Beliefs

It's up to us to recognize that we can't lead a world that we don't understand, and we can't understand the world if we fail to comprehend and honor the central role that religion plays in the lives of billions of people.

– John Kerry

Jesus answered, "You say that I am a king. In fact, the reason I was born and came into the world is to testify to the truth. Everyone on the side of truth listens to me." "What is truth?" retorted Pilate.

– John 18:37-38

It was not expected to work out this way. According to the Secularization Thesis—a social theory arguing that as a culture modernizes and education spreads, religion will die—everything should be different by now.[181] Education is up so faith should be down. But it's not. Just the opposite.

As *Guardian* columnist Giles Fraser wrote recently, the year that Friedrich Nietzsche—the philosopher famous for stating "God is dead!"—himself died (1900), Africa was home to eight million followers of Christ. Today, the number is over three hundred million, and the Christian church is spreading like a brush fire.[182] Islam is also up, and now plays a much bigger role in world affairs than it did twenty years ago.

Even those who do not participate in organized religion are often quick to claim that they are "spiritual."

Despite earlier forecasts, faith is one of the most important forces on the planet today, and no one who hopes to understand how the future will unfold can afford to ignore its influence.[183]

The Final Glacier

I identified faith as the last glacier for three reasons.

» **Faith always has a major impact on the world.** People are different from other animals. We seek purpose and ascribe ultimate meaning to something. It might be a god (classically understood). But it also could be money, sex, or the Chicago Cubs. Everyone has a first love, and that love shapes each person's life for good or ill. The dominant beliefs of the people shape cultures, which in turn impact the world.

» **Faith will shape the future.** Faith is not going away. Individual religions and philosophies may wane, but humanity's need for meaning and purpose is not about to change.

» **Our personal futures pivot around our faith.** The third reason I am highlighting faith is because your ability to navigate the future will be determined in part by your most deeply held beliefs.

This chapter unfolds around five claims: *Everybody Believes in Something; These Beliefs Shape Everything; This Is a Confusing Moment; Five Big Options Remain;* and *A Lot Depends on How This Plays Out.*

Before we move on, let me restate and expand on what I said at the end of the previous chapter. I am using the term "faith" in a broad and generic sense. It includes organized religions, but it is not limited to them. It also includes personal ideologies, animating ideals, and ultimate convictions. In light of this, I will use the terms "faith" and "belief" and "worldview" interchangeably. So now the claims.

One: Everyone Believes in Something

Many believe that only those who join an organized religion exercise faith. This is not accurate. We all have faith. I have faith. You have faith. Even Richard Dawkins—the prominent atheist who denounces faith—has faith. Each of us accepts things we cannot prove.

I realize these are fighting words. Many insist that they are above such nonsense. They believe that they do not believe. "I'm a scientist. My views are grounded in reason and facts alone."

These claims make philosophers roll their eyes. They note that science rests on a number of assumptions that cannot be proven, such as the idea that thoughts reflect reality or the belief that the laws of nature are uniform.

More significantly, you cannot use natural methods to disprove the supernatural. Three hundred years ago, in *A Critique of Pure Reason* and subsequent work, German philosopher Immanuel Kant developed a philosophical basis to argue that science could not prove that there was not a spiritual dimension beyond the realm of science. In light of this, we are free to believe that what we see is all there is, but we should realize that we cannot prove it.[184]

Or to state this more forcefully, we can affirm Carl Sagan's claim that "[t]he Cosmos is all that is or ever was or ever will be,"[185] but we cannot prove it. Sagan's quote is nothing more than a naturalist's statement of faith.[186,187]

Examining your worldview. The idea that we all accept a set of assumptions that we cannot prove is developed in a number of fields. Philosophers call these assumptions our "pre-reflective" commitments, social scientists refer to them as our "plausibility structures," and others discuss our "mental models of the world." I will use the term "worldview." I define worldview as a set of basic assumptions about reality that are accepted without proof, which color people's perception about everything.[188]

Think of your worldview as a pair of glasses that brings the world into focus. Everything we know is filtered through these glasses. Yet remarkably, until we talk with someone wearing different glasses, we are unaware that we have any glasses on.

And realizing that we are wearing glasses is not the first thing we think when we meet someone with a different worldview. Our first thought is that the person is crazy, wrong and weird.[189]

A few years ago I had a run-in with an associate dean who did not believe she was wearing glasses. I had just given a lecture at her university when she approached me to say that in the future she expected me to remain more "intellectually neutral."

When I asked her what she meant, she said, "You cannot suggest that one idea, religion, philosophy, worldview, or approach to life is better than any other."

I'd heard this before and knew that I wasn't likely to get very far, but since I wasn't inclined to accept what she was asking, I replied, "You realize that you're violating your own principles, don't you? What you've just said is a view, and you are claiming that it is better than my view. You are not being neutral. You are doing the very thing you are telling me not to do."

"No, I am not," she said. "My views are not views. I am being unbiased and neutral. I am saying that all views are equal."

"Right, but that is a view," I said. "And you are claiming that it is right. I agree that everyone has a right to their view, and I also think that we should protect people's right to believe what they choose. But that is different. The claim that all views are right is a view, and it isn't even a very good one. In fact, it's nonsense. It's like saying, 'I have a round square,' or 'I always lie.' Your view collapses as soon as we push on it. In order for every view to be right, those who say "every view is right" and those who say "not every view is right" are both right. And that is absurd. What you are suggesting violates the "law of noncontradiction."[190] You are advocating a form of relativism. It's based on the idea that every idea has value. I am arguing that every person has value but not every idea does. Some are better than others."

Her response? "Mr. Woodruff, if you are going to be disagreeable, I'm not going to let you speak to the students anymore. You have to agree to be neutral."

The conversation went on a bit longer, but she didn't budge.[191] She could not see her own set of assumptions. From her perspective, my views were shaped by unproven beliefs but hers were not.

It's hard for us to see our own thinking, our own worldviews, objectively, but it can be done. In fact, there is a branch of philosophy dedicated to this. It's called "epistemology" and it focuses on helping people examine why people believe what they believe.[192]

Before we move on, I should say two things. First, epistemology is not for wimps.[193] Thinking about thinking is difficult, and thinking about our own thinking is disorienting. But it's important to do in order to live well, and it's necessary to do if we want to understand the world. Second, in spite of what you are thinking, this isn't a rabbit trail. I imagine that some of you are wondering, *Where is this headed? Glasses? Epistemology? Thinking about thinking? Give me a break. This does not relate to the future.*

But it does. Please trust me and keep reading.

The seven questions: One of the ways philosophers help people take their glasses off is to ask provocative sets of questions. The set of "The Seven Questions" is one such formulation, which you are familiar with whether you've referred to them by that name or not.[194] Everyone has working answers to these questions, and their answers shape their lives in significant ways.

The questions are:

1. ***What is the most important thing?*** What is of greatest value? What is the highest good? By most definitions, your answer to this question is your god.

2. ***Who am I?*** Am I an eternal being, the temporary pinnacle of the evolutionary process, or something else?

3. ***Where did I come from?*** Do I owe my existence to fate, the hand of God, the accidents of evolution? All of the above?

4. ***What is expected of me?*** Is there a purpose for my life—an assignment? Or am I free to make up my own?

5. ***What happens when I die?*** Is death the end, or do I live on in some way?

6. ***How do I determine right from wrong?*** What is the basis of my morality?

7. ***How do I know what I know?*** Where do I go to answer the first six questions?

Since Netflix came out, not everyone has found time for silly questions about the meaning of life. This doesn't mean they do not have answers to the big seven. We all do. We hold views about life, meaning, purpose, and god that shape everything we think, do, and say. We are all theologians and philosophers.

The question is not whether we have answers to The Seven Questions. The question is, are our answers any good?[195]

»«

A brief time out. If you've not gotten around to reviewing your answers to the Big Seven, put that on your to-do list. What you think matters. And what you think about ultimate issues matters a lot. People who believe that they were made for a higher purpose, are accountable to God, and will live forever—such people live differently than the those who think that "what we see is all we get."

I am not saying that you should believe in God because it will make you a better person. That might happen, but it's not a reason to believe. Nor am I saying that you should believe because faith will help you sleep through the night. Again, that might happen, but it's not a good reason to believe. C.S. Lewis wrote, "I didn't go to religion to make me happy. I always knew a bottle of Port would do that."[196]

Of the various directions your answers to The Seven Questions might lead you, the direction of being a Christ-follower is by no means the easiest. Christian faith rises or falls on Jesus Christ. Either he is God, died in our place, and rose from the dead, or he didn't. (At one level it's a history question.) And if he did, he's not offering a bottle of Port (or any other mind numbing or distracting diversion). But more on that later.

The point is, rather than seek ease or pleasure, we need to seek truth and follow it where it leads.

Two: Beliefs Shape Everything

My first point was that we all have a set of starting assumptions. My second point is that if we are consistent, what we believe shapes everything, which in turn shapes the future.

This claim is not limited to individuals. It is also true for entire cultures, which is obvious if we look around. Atheism leads to a different culture than either Islam or Hinduism does, which is why life in Soviet Russia was quite different than life in an Islamic Iran or a Hindu India. Of course other variables also shape culture, but faith's influence is huge.

This point is not lost on politicians. After stepping down as Prime Minister of England, one of the very first things Tony Blair did was teach a course at Yale entitled, "Globalization and Faith." He did this because he believes these are the most important forces shaping the world today. And fifteen years ago Chinese officials established several hundred centers for the study of Christianity in their country. They did not do this to promote faith in general or Christianity in particular, but because they became convinced that the easiest way to fuel honesty and drive economic prosperity was to persuade people to adopt a Christian worldview.[197]

Of course we do not have to look outside our country to see this. John Adams, the second president of the United States, made a related statement two centuries earlier:

> ...we have no government armed with power capable of contending with human passions unbridled by morality and religion. ... Our constitution was made only for a moral and religious people. It is wholly inadequate to the government of any other.[198]

What we believe matters. What individuals believe about ultimate issues—morality, God, purpose, calling, eternal life—shapes their future. The collective view of the people in a country shapes that country. All of this shapes tomorrow.

Three: This Is a Confusing Moment

If you grew up in the fifties in the United States, it was possible to think that everyone saw things the way you did. After all, most of the people you knew shared the same worldview. But as the planet shrunk, things changed. We started bumping into people whose starting assumptions were different than ours. In fact, for the last twenty years we've been living in a potpourri of ideologies and beliefs.

The removal of the tall walls that used to separate cultures is one of the reasons things grew more confusing. But it was hardly the only one. Let me list three more:

A. We no longer agree on how to define truth. One hundred and fifty years ago most people based their worldview on revelation—which means they answered The Seven Questions by looking to a book (such as the Bible). After the Scientific Revolution, a number of people shifted from revelation to reason. More recently, some have shifted from reason to feelings, and some are claiming that there is a shift occurring beyond that.

Another way to talk about this is to say that one hundred years ago we moved from being a traditional culture to being a modern one, and that in the last twenty years a number of people have transitioned from modern to postmodern, and there is talk that some are now embracing a post-postmodern mindset.[19]

This means that the person standing next to you in the checkout line at Walgreens may define truth differently than you do.

B. We are now expected to act as if whatever someone believes is true, is true. Sincerity recently became an acceptable test of truth. That means, if someone tells you that they really believe something, you are expected to affirm it, no matter how odd their view may seem to you.

In fact, the definition of the word tolerance has also changed from "treating those you disagree with in a civil way," to "affirming whatever someone else affirms." Which means, if you do not affirm someone's sincere beliefs you are often labeled intolerant.[200]

C. Many people now customize their views. It used to be that people in the United States selected from one of two options: Christianity or atheism. There were a few nuances—people needed to choose between Protestant and Roman Catholic and also select their level of commitment. But those were essentially the choices.[201]

Starting in the mid-sixties a few Eastern views were added (e.g., Buddhism and Transcendental Meditation). In the mid-seventies New Age thinking became popular. But that was about it.

Back then people picked one of the established options and went with it. It never occurred to them that they could make up their own view, but that is what has been happening more recently.

The rise of individualism, combined with the elevation of feelings, has led many to design their own religion. Few set out to do this. In fact, most do not even realize that this is what they have done. But in a way our grandparents would have found confusing and comical, many now assemble a worldview in much the same way they order lunch at Subway. "I'll take three parts belief in God, one-part reincarnation, and one-part capitalism. Stir in a dash of Green Bay Packers, some environmentalism, and Dr. Phil. Then sprinkle on a little Ayn Rand. Oh, and I want that on six-inch wheat."[202]

Never mind that what I cobbled together doesn't make any sense. It's what feels right to me.

In his 1985 book, *Habits of the Heart,* sociologist Robert Bellah discusses an early observation of this—he calls it "salad-bar" spirituality—quoting a young nurse named Sheila:[203]

> I believe in God. I'm not a religious fanatic. I can't remember the last time I went to church. My faith has carried me a long way. It's Sheilaism. Just my own little voice... It's just try to love yourself and be gentle with yourself. You know, I guess, take care of each other. I think [God] would want us to take care of each other.

The term "Sheilaism" entered America's lexicon as a placeholder for tens of millions of personal faiths, assembled by people who made up their own religion and now act as if it's true.

»«

Remarkably, there are other issues adding to the chaos, such as the fact that many people are gravitating towards more extreme positions than in the recent past—but you get the point. This is a confusing moment. We have been living in a world with competing worldviews and no agreement on how to determine which is true.[204]

Indeed, we do not even agree on what "true" means. It's as if we are competing in a game in which everyone keeps changing teams and no one knows the rules or how to keep score.

Four: The Options Remaining

I opened by arguing that we all operate on the basis of a worldview. I then stated that these worldviews shape everything before claiming that the last twenty years have been a jumbled mess. We now move to the fourth point.

It turns out that some things have been sorting themselves out. Just as Wal-Mart gobbled up mom-and-pop retailers, a handful of worldviews squeezed out the bit players. Today five worldviews are prominent: Secularism, Buddhism, Islam,

Sheilaism, and Christianity. Most people fall into one of these five camps.

Secularism. Many confuse secularism with atheism. This is understandable because many secular people are atheists, but they are not identical. Technically, secularism argues that society should operate independently of religious influence. The associate dean I argued with advocated secularism. Secularism reached its likely zenith when atheism was the official policy of both the Soviet and Chinese communist governments. Now:

» The best current guess is that about fifteen percent of the United States embraces secularism, though in recent years the number has been declining. Two reasons are cited for the decline: secularism has fallen out of favor with philosophers and secular people are not having many children.

» Though it is declining, secularism is not about to die. It remains popular in Europe, Manhattan, Harvard's faculty lounge, and in the editorial offices of *The New York Times.* It also colors the views of many who do not view themselves as secular, including a growing number of "Nones."[205,206]

Buddhism. Though Hinduism is the largest religion to emerge in the East, the caste system has not gained traction outside of India—but Buddhism has. Sometimes called a religion and sometimes a philosophy, Buddhism was founded twenty-five-hundred years ago in India:

» It encompasses a variety of beliefs and practices based on the teachings of "the Buddha," or "the Enlightened One." Its primary focus is on ways to understand and resolve suffering.

» An estimated five hundred million people follow one of the many varieties of Buddhism. As the world grows more frenetic, many are turning to Buddhism and its practices for help. As a result, it has been increasingly "white and western" for years.[207]

Islam. Before 9/11, few Americans knew much about Islam. That changed quickly. Unfortunately, much of what was learned has focused on the political, military, and terrorist activities of the Islamic extremists, which is misleading. Islam was founded by the prophet Mohammad, a seventh century merchant:

» Mohammad claimed that the angel Gabriel visited him numerous times over the course of twenty-three years. During those visits the angel revealed the words of Allah (the Arabic word for God). These dictated revelations compose the Qur'an Islam's holy book.

» Around 1.7 billion people currently embrace the teachings of Muhammad, and it is the fastest growing of the final five—mostly in North Africa and Europe. "Mohammad" was the second most popular boy's name in London last year, and some predict that Europe will have a Muslim majority by the end of the twenty-first century.[208]

Sheilaism. As noted earlier, Sheilaism, coined by Bellah in his research for *Habits of the Heart,* is a term for cafeteria religions, the beliefs of people who blend strands of multiple religions together, generally without rigorous thought.

Though individuals report assembling their set of beliefs according to an inner sense of truth, most practicing Sheilaism adhere to a number of the same tenants:

» People are good , inherently; in turn, their highest good is their freedom and happiness as individuals.

» Any received wisdom, traditions, or religions that restrict individual freedom and happiness are bad.

» The world on the whole will improve as the scope of individual freedoms grow.

» Tolerance is the primary social ethic; any deviation from it is dangerous and should not be tolerated.[210]

Christianity. Christianity emerged out of Judaism two thousand years ago when Jesus of Nazareth claimed to be the promised Messiah (i.e., Savior of the world). The Christian faith is based in part on his example, teaching and death, but it is grounded in the belief that he defeated death and rose from the dead three days after his crucifixion.

» Approximately two billion people identify as Christians (Roman Catholic, Orthodox, and Protestant), making it the largest worldview. The "church" that encompasses these is the most geographically and ethnically diverse movement in the history of the world.

» In spite of its size and influence, many misunderstand it, assuming that Christianity teaches that God wants people to be good and that good people go to heaven when they die.[211]

The Bible actually teaches that people are broken (sinful), and are therefore unable to be good enough to merit God's favor. Instead of humankind reaching up to God, God reached down to us by sending his Son, Jesus, as an intermediary. Christ's death is understood to be a sacrificial act of atonement designed to reconcile people with God, and with each other.[212]

Unlike Islam, which is growing because of high birth rates—Muslim wives have, on average, ~3 children each, a rate far higher than any other group—Christianity is growing mostly through conversion, especially in the global South and East.[213]

The Church Moves South and East.

–

There are now, as Rick Warren observes, more Presbyterians in Ghana than Scotland, more Baptists in Nagaland, India than the American South, and more Anglicans in Kenya (or Uganda, Rwanda, or Nigeria) than in England. Any given Sunday, more Christians attend church in China than all of Europe combined.[209]

Five: A Lot Depends on How Things Play Out

It is now time to look ahead. So far I have argued that everybody believes in something; these beliefs shape everything, and they have been competing against one another. I also noted that five major beliefs remain. We now ask, "Where are we headed?"

It turns out that a lot depends on how things resolve and, unlike the previous three glaciers, there is not a lot of agreement on what will happen next. Broadly speaking, two scenarios are possible.

First Scenario: One of the five worldviews will overtake the other four. In various ways, each of the five remaining views make claims to this end. For instance,

» Some secularists still argue that religion will die, or completely privatize, and that secularism will prevail.

» Radical Muslims are open about their goal of expanding Islamic Law over every part of the globe.

» Christians argue that history is heading towards a day when Jesus is universally acknowledged as King.

Depending on where you tune your radio dial, you can hear people advancing these theories and others like them.[214]

Second Scenario: The five worldviews will remain locked in an uneasy truce. The more common view among futurists and geopolitical strategists is that the five worldviews will all roll forward, locked in an ongoing tug of war. One or two may die, and one or two may grow larger, but no one view will prevail.

Which scenario is more likely? I expect the second scenario is more likely in the near term, but as always, much depends on what happens in China (where the church is currently growing rapidly) and India (where militant Hinduism is on the rise). Since these two countries make up nearly one-half of the world's population, the path they take is important.

From my perspective, we can also be reasonably certain of several things:

» ***Europe will become religious again.*** Though this may surprise some, both Evangelical Christianity and Islam are growing there. Meanwhile, secularism is in significant decline.

» ***Islam's growth will slow.*** The Muslim birthrate will remain higher than that of most of Europe, so the percentage of Muslims on that continent will grow. Additionally, Islam may continue to grow faster than Christianity in the near term. However, Islam's growth has been fueled almost exclusively by the large number of children being born to Muslim women, and that now is slowing down.

» ***Sheliaism will lose market share.*** When life becomes more difficult—as I believe it will--many will find that the generic spirituality they cobbled together does not provide the level of support they need.

Perhaps more can be said, but in order to, I believe we need to know how the answers to two critical questions will unfold:

» ***Which path will Islam take?*** Moderate and radical Muslims are currently engaged in a struggle over the direction of Islam. I believe the moderates will eventually prevail, but more than a few think I am wrong. For the record, even if I am right, I believe that, Muslim radicals will be a disruptive (and terrorizing) force into the foreseeable future.[215]

» ***Will the West embrace principled pluralism?*** At the moment, many in positions of influence within the government and media ascribe to the views of the Associate Dean. They want a neutral public square and many are intent to do virtually anything they can to silence those who disagree with them. Faith will be allowed, but only if it is private.

»«

One of the five worldviews will overtake the other four, or will they remain locked in an uneasy truce? What courses will secularism, Islam, Christianity, and the pick-and-choose personal faiths take in the future, in America, Europe, and around the world? Many things we can engage only as spectators, but other things are fully in our control. And so now we will turn to each of our next steps, and my final thoughts.

My Discussion With an Imam.

–

Several months ago I spent four hours with a young imam. Born and educated (through college) in the United States, he had recently returned from five years of training in the Middle East. We had a wonderful discussion, but he took offense when I asked him whether "moderate" Muslims would prevail over radicals. After noting that seventy thousand imams in India had recently condemned ISIS, he argued that: 1) no one who is serious about their faith wants to be called "moderate;" and 2) that radical jihadists are not true Muslims.

Next Steps and Final Thoughts

The man who has no inner life is the slave of his surroundings... – Henri Frederic Amiel

I used to think that top environmental problems were biodiversity loss, ecosystem collapse and climate change. I thought that thirty years of good science could address these problems. I was wrong. The top environmental problems are selfishness, greed and apathy, and to deal with these we need a cultural and spiritual transformation. And we scientists don't know how to do that. – Gus Speth

The simple believe anything, but the prudent give thought to their steps. – Proverbs 14:15

So now what?

We've water-skied over an ocean of information and I'm guessing you're feeling a bit overwhelmed. I am. And that is the general tone of those who read early drafts of this book.

More what? Volatility? Nanotechnology? Swirling ideologies?

Some wanted to race out and do something. Anything. Buy gold, Plant a garden. Move to Montana, etc. A few started complaining, "Everything's happening too fast. It's the government's fault. I blame Steve Jobs." (Or the liberals, the conservatives, the UN, Walmart, etc.)

But most just said, "Wow," and reported feeling unsettled. Then they asked, "So now what? What am I supposed to do? If you are right about the future, what does that mean for me?"

Let's set the "If you're right..." part aside. This book has focused on the major trends. I distilled ten thousand pages of books, articles, marketing forecasts, trend projections, and reports down to less than one hundred pages (one hundred and fifty if you count the notes at the end). I did this by focusing on the areas where most people agree, and dismissing the outliers. I'm sure I'm wrong on some specifics, and there is always the chance that one of the "Monsters Under the Bed" will jump out. But I believe the trajectory mapped out here is a pretty safe bet. Virtual reality may turn out to be a bust, but technology is going to shake us like we're rag dolls. Your job may be safe, but jobs will go away. Globalization may affect us less than the federal debt, but globalization is a big deal. I'm willing to concede the specifics, but I stand by the main ideas.

So what do we do?

The Request for a Three-Point Memo

A friend who's been listening to me process all of this over the last two years recently said, "I don't have time to read your book. When you're done writing it, reduce it down to three bullet points and send me an email."

I was tempted to reply, "I've got it down to one point! The problem with the world is people who do not have time to read a short book because they are too busy playing Pokemon Go."

But I didn't. I decided to meet him partway. I reduced the book down to four brief points.

A Lot of Things Are Going Well

There will always be people yelling that the sky is falling. One day it will. But I come away from my study of the future more optimistic than I went in. Studying the past—which I will remind you is a big part of how we look ahead—reminded me that humans are resilient and that worry is overrated.

The glass is more than half full. There is much to celebrate. And I believe we will find solutions to many of the problems heading our way.

There is Rough Weather Ahead

But not all is well and it's foolish to tell ourselves otherwise. Neither the prophets of gloom nor the utopian idealists are doing us any favors.[216] We need to set aside our biases and face the facts, whatever they are.

And what the facts suggest is that things are about to get even more "interesting" than they have been. We're on a roller coaster. It's picking up speed and there are rapid descents, hairpin turns, and backwards loops just ahead. I believe you can survive this ride, but you need to strap in and hold on.

We Need Wisdom

Many things are getting better, but tomorrow will not be kind to everyone. The one-two punch of globalization and accelerating technology will strengthen the hand of some but disrupt many. The sexual revolution is going to send us a bill that is much higher than most people realize. It's not clear how long it will take us to learn how to get along on a smaller planet. Or if we even can.

And those are just a few of the challenges. Tomorrow is going to require different things from people than today. One of them is self-leadership. You will need a stronger inner world to withstand the swifter currents and increasing velocity of tomorrow.

Nothing will be easier in the future than being busy. Few things will be less common than living thoughtfully. Most people are not prepared for what is headed our way. Those reading a book like this can likely manage massive amounts of data, and even hold their own when it comes to knowledge. But most of us lack the wisdom that flows out of a life well lived and a vibrant relationship with God.

It doesn't matter how well things are going "out there" if our heart is a dark, jumbled mess. To live well in the future we need to latch on to timeless truth. I fear tomorrow will make its greatest demands of us exactly where we are weakest.

Slow Is the New Fast.

–

There will be more options but few of us can do more. Moving faster is no longer a viable strategy. Those who successfully navigate the future will do so not because they can operate at the speed of technology, but because they slow down and reflect.

A Pastoral Response

Let me stop reporting for a moment and step into the role of pastor. So far I've limited most of my comments to advocating things like: work hard to remain employable, stay proficient on the tech front, be prepared for the greater hurry and increased competition globalization is sending our way.

But there is an entirely different avenue that needs to be considered. Preparing for the future also means identifying what it means to live well, cultivate good friends, and find lasting purpose.

It also means preparing to be knocked down. At some point we all run into problems we cannot overcome. Two years ago I suffered a rare brain injury. It was thunder from a blue sky. On Thursday I had a physical and was given a clean bill of health. On Friday I had a spontaneous cerebral artery dissection (SCAD) and ended up in a neuro-ICU ward under the care of four teams of doctors. It was four months before I could stand up and one year until I could drive. Had it not been for the care of my family and friends and an abiding sense of God's love and care, I am not sure where I'd be today.

Let me be as candid as I dare. I've spent the last eighteen months reading about the future. I believe there is great value in looking ahead, but in the end the only thing I can be certain of about your future is that at some point you will face challenges you cannot rise above. That means part of preparing for tomorrow must include a set of things few people talk about today, such as cultivating a rich relationship with God, preparing to die, and developing great friends.

We Need Friends

Social Media notwithstanding, technology has an isolating effect. It enables us to fly solo more often than in the past. And our wealth allows us to pay people to do things that we would have asked our neighbors to help us with twenty years ago. The net effect is that we may have a wide network but we lack good friends.

This may make life easier, but it does not make it better. It also leaves us without the kind of safety net we need when it matters.

When climate change activist Bill McKibben was asked how to prepare for the trials he is convinced are headed our way, his advice was simple: live "anyplace with a strong community."[217] McKibben argues that, in the end, our ability to navigate future challenges will pivot on help from a neighbor.

As a pastor I visit people in the hospital. On some occasions I step into a packed room. Friends, neighbors, and a half a dozen people from their small group Bible Study from church have all turned out. It's quickly obvious that this person has an army of practical and emotional support. I leave feeling hopeful.

On other occasions I enter a room and realize, with the exception of the medical staff, I will be the only one showing up today. In those situations I expect the worst.

The future will be a challenge for those with good friends and a solid faith. It will be immeasurably harder (and more frightening) for those without.

»«

Peter Drucker argued that the best way to predict the future is to create it. No one is better positioned to shape your future than you are. Tomorrow has much to offer to those who prepare. And the most important prep work to be done is on your own soul. My hope is that you'll pass this book to a neighbor and then invite them over for dessert to talk about the ideas it introduces.

Notes

Chapter One, *Why Study the Future?*

[1] Gail Collins, Angela Creager, et al. "What Was the Worst Prediction of All Time?," *The Atlantic*, May 2015.

[2] The caption in the *Technology Review* ran under a picture of former astronaut Buzz Aldrin, now eighty-two years old. It is a play on PayPal founder, Peter Thiel, et al.'s "We wanted flying cars. Instead we got 140 characters." Jason Pontin, "Why We Can't Solve Big Problems: Has Technology Failed Us?," *MIT Technology Review*, October 24, 2012. See also Virginia Postrel, "No Flying Cars, But the Future Is Bright," *Bloomberg*, December 16, 2012.

[3] Malthus' views were widely disseminated and later hijacked by some prominent public intellectuals—such as George Bernard Shaw and D.H. Lawrence—to support eugenics and forced sterilization. Shaw is quoted as having said, "the majority of men at present in Europe have no business to be alive." In 1908, Lawrence wrote in a later published letter:

> If I had my way, I would build a lethal chamber as big as the Crystal Palace, with a military band playing softly, and a Cinematograph working brightly; then I'd go out in the back streets and main streets and bring them in, all the sick, the halt, and the maimed; I would lead them gently, and they would smile me a weary thanks; and the band would softly bubble out the "Hallelujah Chorus."

See Jonah Goldberg, "Let It Grow," *National Review*, October 18, 2006. The letter collection of the Lawrence quote, with citation, can be found at *Wikiquote*.

[4] On the CIA's ignoring the religious dimension of the radical change in Iran, see Robert Jarvis' analysis (studied at the CIA), which concluded that agents were studying everything there except religion, because they were quite confident that "the religious dimension... was... an anachronism." See Torrey Froscher, "[Discussion of] Robert L. Jervis "Why Intelligence Fails: Lessons from the Iranian Revolution and the Iraq War,"" *Studies in Intelligence*, Vol. 54, No. 3, December 10, 2010 (Langley, VA: CIA Center for the Study of Intelligence, Intelligence in Public Literature). The book described is R.L. Jervis, *Why Intelligence Fails: Lessons from the Iranian Revolution and the Iraq War*, Cornell Studies in Security Affairs (Ithaca, NY: Cornell University Press, 2010).

[5] As Thomas Long describes,

> The 1939 New York World's Fair was dubbed "the world of tomorrow," and millions of visitors sported buttons proclaiming, "I Have Seen the Future!" The fair promised to rocket people out of the gloom of the Depression into the wonders of 1960, but what visitors actually saw were things like Tomorrow Town, a neighborhood that looked like suburban Levittown, a futuristic "smell-o-vision" movie theater that never materialized, and a Westinghouse-sponsored dishwashing contest between Mrs. Drudge, with dishpan hands, and Mrs. Modern, who sported a cocktail dress and effortlessly loaded dishes into a spiffy electric dishwasher."

As Long notes, in their optimism, these visionaries for the world of tomorrow missed the ongoing rise of the Third Reich, conflict with which would soon begin. See Thomas G. Long, "Future Fatigue," *The Christian Century*, June 21, 2012.

[6] Today there is a three-hundred-billion dollar industry marching under the banner of Futuring (future studies, forecasting, et al.). And though there are hacks who overpromise (basing their predictions on little more than tea leaves and sheep entrails), many futurists are doing the hard work of turning forecasting into a science. As a result, we are learning what to pay attention to, how to eliminate projection biases, how to craft teams that can look five years out and more.

[7] Tim Challies, "Thinking About Change," *TableTalk* (Sanford, FL: Ligonier Ministries), January 1, 2016.

[8] Oliver O'Donovan has warned that "those who do not make an effort to read their times in a disciplined way read them all the same, but with a narrow and parochial prejudice." He further insists that wisdom about our times requires a detachment and courage, "to attend especially not to those features which strike our contemporaries as controversial, but to those which would have astonished an onlooker from the past but which seem to us too obvious to question." See Ken Myers, "Contours of Culture: From Heavenly Harmony, the Wonder of His Works," *Touchstone*, Vol. 27, No. 6, March/April 2014.

[9] Paul Kinsinger, "Adaptive Leadership for the VUCA World," *Thunderbird Magazine* (online), July 21, 2016.

[10] The First Industrial Revolution used water and steam power to mechanize production. The Second used electric power to create mass production. The Third used electronics and information technology to automate production, a digital revolution that has been occurring since the middle of the last century. Now a Fourth Industrial Revolution (sometimes referred to as Industry 4.0) may be building on the Third. It is described as a fusion of technologies that blur the lines between digital, physical, and biological areas of technology.

[11] I am reporting Collins' comments at a private dinner, as shared by Bob Buford, "Am I Frightened or Encouraged?," *Leadership Network*, August 18, 2010. Regarding the so-called Doomsday Clock, see BAS Staff, "Doomsday Clock: Timeline," *Bulletin of Atomic Scientists* (online), July 13, 2016. For an account of the views of Hawking, Musk, and Bill Gates on AI, see Michael Sainato, "Stephen Hawking, Elon Musk, and Bill

Gates Warn About Artificial Intelligence," *Observer.com* (online), August 19, 2015. For more on Hawking's dire predictions, see James Gerken, "Stephen Hawking Predicts Humans Won't Last Another 1,000 Years On Earth," *The Huffington Post* (online), April 28, 2015.

[12] It is possible that there is a pessimism bias today because futurists do not want to repeat the mistakes made by their nineteenth century predecessors (i.e., those who promised a world Utopia as they headed, as Albert Mohler describes, towards "Hitler's gas chambers, Stalin's gulags, and Pol Pot's Cambodian killing fields"). With few exceptions (e.g., Nietzsche), nineteenth century forecasters had bought into so much Enlightenment optimism that they missed the trouble that was headed their way. Albert Mohler discusses Francis Fukuyama's 1992 book, *The End of History and the Last Man* with regard to this subject. Mohler notes that Fukuyama explains how "[t]he nineteenth century's humanistic faith in inevitable moral progress was destroyed on the battlefields of two cataclysmic world wars," and how "[h]istory seems to point not to a golden age of moral progress and enlightenment, but toward an age of unspeakable cruelty backed by technological developments that would stagger the moral imagination." Per Fukuyama, "The twentieth century, it is safe to say, has made us all into deep historical pessimists." See Albert Mohler, "Christ the Victor," *TableTalk* (Sanford, FL: Ligonier Ministries), March 1, 2005, and Francis Fukuyama, *The End of History and the Last Man* (London: Penguin Adult, 1992), p. 3.

[13] Among those who contend that predictions of economic gloom and resource depletion are wrong because they fail to adequately account for the ingenuity of business people and scientists to come up with innovative solutions are Matt Ridley (author of the *The Rational Optimist*, see below) and George Gilder (founder of the Discovery Institute).

Chapter Two, *Many Things Are Getting Better*

[14] Kevin Kelly is author of works that include *The Inevitable: Understanding the 12 Technological Forces That Will Shape Our Future* (London; Penguin, 2016). In this conversation with him, I countered that the new jobs sounded good, but that they assumed a pretty high education and skill level, which didn't seem within reach of either the laborers who are generally among the first to be replaced, nor the kids growing up in urban poverty. Kelly agreed, noting that transition periods—like the one we are entering—are ugly.

[15] This conversation took place in around our participation in the Q Conference sponsored by *Qideas.org* in April 2013, parts of which addressed artificial intelligence and other technology subjects.

[16] The list of things that were going to sink us was long and foreboding, including most of the faculty's near certainty that Reagan was going to crash the economy or start a nuclear war with the USSR.

[17] Dan Alban, "Put Your Money Where Your Mouth Is," *The Harvard Law Record*, October 6, 2005.

[18] Julian L. Simon and Paul R. Ehrlich engaged in the wager in 1980 that was documented in the *Social Science Quarterly*. However, the debate began in public, after Ehrlich's repeated appearance on *The Tonight Show with Johnny Carson* (which would eventually number in the 20s). As Ed Regis noted in *Wired* in 1997, Ehrlich's public claims included the statement, "If I were a gambler, I would take even money that England will not exist in the year 2000." In the wager, Simon focused on the issue of price and resource scarcity. Specifically, Simon made "a public offer to stake US$10,000... on [his] belief that the cost of non-government-controlled raw materials (including grain and oil) will not rise in the long run." He then offered Ehrlich the challenge to choose 1) a raw material, and 2) a date more than a year out from the date of the wager. He would bet that the price of the raw material would decrease (in inflation-adjusted terms). Ehrlich chose metal commodities—copper, chromium, nickel, tin, and tungsten—and the wager was formalized on September 29, 1980 (with the same date, a decade later, in 1990, to be the payoff date). Ehrlich lost the wager; all five materials listed declined in price from 1980 to 1990. As Dan Alban notes, "Simon offered a follow-up wager for $20,000, [which] Ehrlich refused but [he] still maintained that resources were becoming increasingly scarce." For the opening quotations, see Ed Regis, "The Doomslayer," *Wired*, Issue 5.02, February 1997, p. 5, and for the last one, see Alban, "Put Your Money..." (last note). For further background, see David Kestenbaum, "A Bet, Five Metals And The Future Of The Planet," *NPR's Morning Edition*, January 2, 2014, and the Economist Staff, "Innovation in History: Getting Better All the Time [Book Review, The Rational Optimist: How Prosperity Evolves. By Matt Ridley]," *The Economist*, May 13, 2010.

[19] See Max Roser, "Life Expectancy," *Our World in Data* (online, Oxford: Institute for New Economic Thinking), July 22, 2016. David H. Bailey writes,

> Even most die-hard pessimists acknowledge that people are living longer than in earlier decades and centuries, but it is widely believed that this increase has leveled off in the last decade or two, particularly in North America and Western Europe. Not so. The latest 2011 statistics from the U.S. government's Division of Vital Statistics indicate that life expectancy rose for the tenth consecutive year, to an average of 78.2 years, up from 70.8 years in 1970.... The principal blight on the U.S. picture is that its ranking among developed countries has slipped from 20th in 1987 to 35th today..., with even greater slippage among poor women.... Worldwide average life expectancy reached 71.5 years in 2013, up from just 65 years as recently as 1990, based on the latest data from 188 nations....

See D. H. Bailey, "Is Modern Society in Decline?, [Section "Other statistics: Where is the decline?," Subsection 10, "Life expectancy."] *Science Meets Religion* (online), July 21, 2016.

[20] Sociologist Brad Wright notes that United Nations estimates put global poverty in greater decline over the last fifty years than over the previous five hundred. See the text for more from him; Bradley R.E. Wright, *Upsides: Surprising Good News About the State of Our World* (Bloomington, MN: Bethany House, 2011), p. 63. In The Economist's "The World in 2015," Bill Gates is quoted saying,

> In the past twenty-five years, the number of children who die [people who die in childhood] has dropped by a half. In 1990, 12.7m children died. If the rate of death had stayed the same, then the number of children who died last year would have been more than 17m, if you take population growth into account. Instead, it was just over 6m. The number of extremely poor people has been going down at roughly the same rate, with the percentage of young people in the world cut by more than half since 1990.

See Bill Gates, "Great Expectations," *The World in 2015* (print and online, London: The Economist Group, November 20, 2014), p. 93.

[21] David Bailey writes,

> It is widely believed that crime, from minor burglary to serious violent offenses, is spiraling out of control, and is prima facie evidence of societal disintegration and a wholesale breakdown of morality. Yet the facts point in quite the opposite direction. Indeed, the latest U.S. crime data has stunned even the most optimistic of observers. The 2013 violent crime rate was 5.4% lower than in 2012, and the 2013 property crime rate was also 5.4% lower than in 2012, with similar year-to-year declines for at least 15 years. These rates are down by more than a factor of two since peaking in 1994. . . .

See Bailey, "Is Modern Society. . .," [Subsection 1, "Crime."]. The World Health Organization (WHO) estimates that, "[d]uring the period 1990-2002 (for which data are available) global [drinking water] coverage rose by 5 per cent, from 77 to 83 per cent. This means that nearly 1 billion people gained access to improved water sources during this period." See WHO Staff, "Health Through Safe Drinking Water and Basic Sanitation: What's Needed to Reach the Target," *WHO.int* (online, Geneva, CHE: WHO, Water Sanitation and Health, July 22, 2016).

[22] Most recent data from this source is for 2010, see the opening chart, data for "World," at Max Roser, "Literacy," *Our World in Data* (online, Oxford: Institute for New Economic Thinking), July 22, 2016 (where mousing over data points give values). See also our Figure 2.1 in the main text, and the surrounding discussion.

[23] See Wright, *Upsides,* p. 129, and Michael Cox and Richard Alm, "By Our Own Bootstraps: Economic Opportunity and the Dynamics of Income Distribution" [Annual Report] (Dallas: Federal Reserve Bank of Dallas, 1995), p. 22 and *passim.*

[24] Greg Easterbrook claims that people living in the middle class in the U.S. today live better than 99.4 percent of all human beings who have ever existed. See Greg Easterbrook, *The Progress Paradox: How Life Gets Better While People Feel Worse* (New York: Random House, 2003), p. 80.

[25] For entrée into the story regarding the United States, see Steven Johnson, "We're Living the Dream; We Just Don't Realize It," *CNN,* November 24, 2012. See also Wright, *Upsides, passim,* and the other references in our chapter two.

[26] Claude S. Fischer and Michael Hout, *Century of Difference: How America Changed in the Last One Hundred Years*, The Russell Sage Foundation Census Series (New York: Russell Sage Foundation, 2006), p. 139f.

[27] Fareed Zakaria, "Excerpt: Zakaria's 'The Post-American World,'" *Newsweek* (online), May 3, 2008. Also appears as "The Rise of the Rest," *FareedZakaria.com,* May 12, 2008. The book from which the excerpt is taken is F. Zakaria, "The Post-American World" (New York: W.W. Norton, 2008).

[28] Another book that deserves mention is *The Skeptical Environmentalist,* by Bjorn Lomborg, a professor of statistics at the University of Aarhus in Denmark. Lomborg's book has created a storm of protests and I am not able to tell how much of it is fair. I am simply noting that against the claim that we are rapidly losing forest cover, Lomborg contends that the depletion stopped some years ago and has been holding steady at thirty percent. Against charges that fifty percent of all species have died out within the last fifty years, Lomborg claims the number is actually 0.08 percent each year (or 4 percent over 50 years). He goes on to note that whales are no longer threatened, that elephant herds are now being culled because their numbers are so high, and that "air pollution is not a new phenomenon that is getting worse and worse, but an old phenomenon that is getting better and better, leaving London clearer than it has been since the Middle Ages." Again, I am not able to tell who is changing the facts to support their opinions. I am simply noting that Lomborg is among those who argue that many things are getting better, and that those things not getting better are getting worse at a slower rate. See Bjørn Lomborg, *The Skeptical Environmentalist: Measuring the Real State of the World* (Cambridge University Press, 2001), p. 164. See Wright, *Upsides,* pp. 179ff, and Anthony Browne, "Recovering Earth," *The Guardian,* June 10, 2001.

[29] Data are graphed from the report by Nicholas Burnett and the Education for All (EFA) Global Monitoring Report Team, *Literacy for Life* (Paris: United Nations Educational, Scientific and Cultural Organization [UNESCO], 2005), p. 165f, 23.

[30] Data are re-graphed from Figure 1 in MMWR Staff, "Achievements in Public Health, 1900-1999: Healthier Mothers and Babies," *Morbidity and Mortality Weekly Report* [MMWR], Vol. 48, No. 38, October 1, 1999 (Atlanta: Centers for Disease Control and Prevention [CDC]), pp. 849-858.

[31] We may not believe every word, but there is some sort of downward cumulative effect over time. After reading enough, we end up believing that we too are being poisoned by the latest thing reported in a far off news report, even though local levels and expertise say there is no scientifically justifiable reason to worry.

[32] Staff of the Curry School, "School Violence Myths: School Survey Hoax," *Curry.Virginia.edu* (Charlottesville: Rector and Visitors of the University of Virginia, 2015).

[33] In the 1980s, noted sociologist Peter Rossi "conducted a rigorous count of the number of homeless people in Chicago and used [that] number to estimate the population nationwide." After his study, Rossi said that a more

accurate number would be 150,000-200,000 (see Wright, *Upsides,* p. 39). At the time, it was not uncommon for people to claim that there were two million homeless people in the United States. On the one hand, this was good news! The problem was far smaller than previously thought. But Rossi faced a lot of opposition from homeless advocates because they feared his report would decrease their funding.

[34] Philly.com Staff, "Greenpeace Just Kidding About Armageddon," *The Washington Post,* June 2, 2006. See also Anthony Browne, "Recovering Earth," and Wright, *Upsides,* p. 181, cited in note 20, above.

[35] Wright, *Upsides,* p. 17.

[36] Wright, *Upsides,* p. 19.

[37] See Wright, *Upsides,* p. 20, who refers to David Whitman, *The Optimism Gap: The I'm Ok–They're Not Syndrome and the Myth of American Decline,* (New York: Walker and Company, 1998), and F. Moghaddam and C. Studer, "The Sky is Falling, But Not on Me: A Cautionary Tale of Illusions of Control, in Four Acts," *Cross-Cultural Research,* Vol. 31, No. 2, pp. 155-167.

[38] Will Rogers is said to have quipped a correction of a past-remembered too rosy, with "Things aren't what they used to be and probably never were." Some say we have a dour disposition about the future because we forget the bad about the past.

[39] Among other things, Wright notes that the view that everything is getting worse makes it hard to prioritize the matters that we actually do need to focus on. Quoting David Whitman, he notes, "False alarms drive out true ones." (Wright, *Upsides,* p. 29).

[40] This is essentially a partial introduction to the chaptering in Wright, which include "upside" coverage of our finances, thought life, physical and emotional health, the state of public safety, our marital and family relationships, and the environment. See Wright, *Upsides,* p. 9 for a roadmap to his treatments of these same topics.

Chapter Three, *Glaciers Reshaping the Globe*

[41] See James Martin, *The Meaning of the Twenty-first Century* (London: Penguin-Riverhead, 2006). John Naisbitt's work was *Megatrends: Ten New Directions Transforming Our Lives* (New York, Warner Books, 1984).

[42] Richard Dobbs, James Manykia, and Jonathan Woetzel, *No Ordinary Disruption: The Four Forces Breaking All the Trends* (New York: McKinsey Global Institute, April 2015), p. 1. According to this report, the four forces that are colliding and transforming the global economy are the rise of emerging markets, the accelerating impact of technology on the natural forces of market competition, an aging world population, and the accelerating flow

of trade, capital, and people.

[43] One example of a good game-changer would be the discovery of free (or very cheap), clean energy.

[44] Hollywood loves world-ending scenarios, consequently every summer delivers at least a few films featuring zombie revolutions, robotic ascensions, and asteroids raining down mayhem from on high. I do not lose sleep over these. But it would be irresponsible to talk about the future without noting that there is a category of deleterious events that could change everything. For a bit more on technology and the cinema, see the notes for chapter four.

[45] Independent Staff, "Bill McGuire: Prophet of doom," *The Independent* (UK), November 8, 2005. See also David Thompson, "Possible Endgames [Book Review, *Surviving Armageddon: Solutions for a Threatened Planet.* By Bill McGuire, Oxford University Press, 2005]," *Chemistry World,* December 2005.

[46] McGuire is particularly exercised by the LaPalma volcano in the Canary Islands, believing that if it erupts, the tsunami (wave) would be over three thousand feet high when it moved through the Caribbean, and at least one hundred feet high when it crashed into the coastlines of Africa, Asia, Europe, and the Eastern seaboard of the United States. Though the theory behind this claim was published in a peer-reviewed journal, it does not enjoy wide support in the scientific community. Other Gee-Gees of note (for GGEs, global geophysical events) include the damage—and perhaps the wave—caused by the impact of an asteroid or comet. See, for instance, Bill McGuire, *Surviving Armageddon* (see last note), and Susan Casey, *The Wave: In the Pursuit of the Rogues, Freaks and Giants of the Ocean* (Toronto: Random House/Doubleday Canada, 2010), p. 116f.

[47] Daniela Hernandez, "Scientific Doomsday: Ways the World Could Actually End," *Wired,* January 2012.

[48] Hernandez, "*Scientific Doomsday.*"

[49] See **a)** USAID Staff, "Fact Sheet: Emerging Pandemic Threats," *USAID.gov* (Washington, D.C.: United States Agency for International Development [USAID], May 24, 2016), and **b)** Gates Foundation Staff, "Preparing for Pandemics: How Systems, Planning and Research Can Protect the World in the Next Outbreak," *GatesFoundation.org* and *The New York Times* (Seattle: Bill & Melinda Gates Foundation, July 22, 2016).

[50] See **a)** Lois Parshley, "When the State Wilts Away," *Bloomberg,* June 9, 2016. The article is subtitled, "In weak nations, environmental stress can tip society into catastrophe." **b)** NAS Staff, *Global Climate Change and Extreme Weather Events: Understanding the Contributions to Infectious Disease Emergence* [Workshop Summary, Institute of Medicine Forum on Microbial Threats]," (Washington, D.C.: National Academies Press, 2008); **c)** EPA Staff, "Climate Change Impacts International," *EPA.gov* (Washington, D.C.: Environmental Protection Agency [EPA], July 22, 2016); and the entertaining lay compilation, **d)** Tia Ghose, "Doomsday: 9 Real Ways the Earth Could End," *LiveScience,* May 30, 2013.

[51] BAS Staff, "Doomsday Clock," cited in chapter one above.

[52] On use of this economic allusion to describe a disaster scenario, and further examples, see **a)** Ghose, "Doomsday," **b)** EPA Staff, "Climate Change Impacts International," and c) Parshley, "When the State Wilts."

[53] Ghose, "Doomsday."

[54] Rawn Shah, "Risks Impacting the World in 2015 and the Next Ten Years [report on the World Economic Forum annual report on Global Risks, 10th Edition, 2015]," *Forbes,* January 15, 2015.

Chapter Four, *The Acceleration of Technology – Part I*

[55] This is the working definition I arrived at for our discussions. Others that reflect the challenge of defining this rapidly changing concept: **a)** Jim Kerstetter, "Why Some Start-Ups Are Called Tech Companies and Others Are Not, *The New York Times,* August 2, 2015; **b)** Robert Angus Buchanan, "Technology: History of Technology [Introduction, Perceptions of Technology, and The 20th century]" *Encyclopædia Britannica* (online), July 21, 2016; **c)** Science for All Staff, "The Nature of Technology," *Project2061.org* and *Science for All Americans* (print and online, Washington, DC and Oxford: American Association for the Advancement of Science and Oxford University Press, July 22, 2016 [1990]); and **d)** Harm-Jan Steenhuis and Erik J. de Bruijn, "High Technology Revisited: Definition and Position," *IEEE International Conference on Management of Innovation and Technology,* 2006.

[56] In his discussion of how driverless cars will change our lives, David Gibson notes that: **a)** we will stop running errands, **b)** parking hassles will disappear, **c)** we will get more exercise, **d)** we will not own a car, **e)** we will be stuck in more traffic, **f)** we will work more, and **g)** we will never stop "driving" (being driven). David K. Gibson, "Seven Ways the Driverless Car Will Change Your Life," *BBC.com,* 21 March 2016. See also Wallace Witkowski, "Why Robots Will Pay Less for Car Insurance Than You Will," *MarketWatch,* June 1, 2016.

[57] Another way self-driving cars will change things is by doing away with city parking structures. Cars will likely park further away, which will change the look and feel of cities.

[58] Those who think long and hard about these kinds of things further argue that societies move through various stages: ***stage one,*** simple tool-making cultures—of which few are left today—in which tools support the culture without fundamentally changing it; ***stage two,*** societies where tools start to disrupt the traditions and structures of the culture; and finally, ***stage three,*** where technology is the culture. It is at this point that we begin to live very different lives—in the words of Thoreau, we "become the tools of [our] tools." Most who write about this argue that the United States went through this transformation some time ago. See,

for instance, Neal Postman, *Technopoly: The Surrender of Culture to Technology* (New York: Knopf Doubleday-Vintage, 2011 [1993]), 21ff. The full quotation and source from Thoreau's *Walden* can be found at *Wikiquote.*

59 I am thinking here of things like the potter's wheel, stirrups, gunpowder, and the steam engine.

60 See the "Mass Use of Inventions" (graphic) of Ray Kurzweil, "The Law of Accelerating Returns [Essay]," *Kurzweilai.net,* March 7, 2001. As one example of how this image has evolved, see Brent Merswolke, "The Internet of Things and Transportation," *Miovision.com,* July 17, 2013.

61 For reasons of space and familiarity rather than importance, we omitted discoveries the many other types of technologies, e.g., *chemtech, biotech,* and *medtech* (i.e., penicillin and asymmetric catalysis, PCR and humanized antibodies, and MRIs and laser surgery). Some compilations of important discoveries across the years, in all fields of technology: **a)** James Fallows, Michelle Alexopoulos, Leslie Berlin, et al., "The 50 Greatest Breakthroughs Since the Wheel, *The Atlantic,* November 2013; **b)** Daniel C. Schlenoff, "What Are the 10 Greatest Inventions of Our Time?," *Scientific American* (online), November 1, 2013 (which also presents opinions of readers from 1913 for comparison); **c)** Mark Lorch, "Five Chemistry Inventions That Enabled the Modern World: What do Pencillin, Polythene and Mexican Yam Have in Common?," *The Guardian,* 5 June 2015; **d)** NAE Staff, "Health Technologies Timeline, *GreatAchievements.org* (Washington, DC: National Academy of Engineering [NAE], 2016); **e)** David Lubertozzi, "Life Since the Double Helix: 60 Years of Evolution in Biotechnology," *Bioradiations.com,* January 14, 2014; and **f)** Buchanan, "Technology," *Encyclopædia Britannica,* cited above.

62 Eric Gastfriend, "90% of All Scientists That Ever Lived Are Alive Today," *FutureofLife.org* (Cambridge, MA: The Future of Life Institute, July 21, 2016). As Gastfriend comments,

> This simple statistic captures the power of the exponential growth in science that has been taking place over the past century. It is attributable to Derek de Solla Price, the father of scientometrics (i.e., the science of studying science), in his 1961 book, *Science Since Babylon* [New Haven, CT: Yale University Press, 1975]. If science is growing exponentially, then the major technological advancements and upheavals of the past two hundred years are only the tip of the iceberg.

For Toffler's noteworthiness in futurology, see James Joseph O'Toole, "Futurology" and related articles, *Encyclopædia Britannica* (online), July 21, 2016. An available edition of his is Alvin Toffler, *Future Shock* (New York: Bantam, 1984 [1970]).

63 You can think of the multiplicative nature of technology's impact in this way. If A = the number of items that are changing, B = the magnitude of the changes, and C = the pace at which these changes are occurring, then the impact of technology can be thought of as the product of these three metrics, Impact = A x B x C. For instance:

- One thousand years ago the impact of technology was small, and so we might represent it as: 1 x 1 x 1 = 1, or perhaps 2 x 1 x 1 = 2.
- After the Scientific Revolution in the sixteenth through nineteenth centuries, the variables changed. Think: 3 x 2 x 2 = 12
- After the 1950s, with the advances underpinning integrated circuits, progress in biotechnology, etc., things started spiking: 5 x 4 x 5 = 100

Today all three variables are growing rapidly: we are experiencing changes at work and at home, the magnitude of these changes is growing in size, and they are coming at us at an accelerating rate. The result is an equation that looks like this: 15 x 10 x 30 = 4,500.

[64] For Ray Kurzweil's thesis, see Kurzweil, "The Law...," p. 1, cited in note 60. See also Edward Cornish, *Futuring: The Exploration of the Future* (Chicago: World Future Society, 2004), p. 12. The book referred to at paragraph opening is Ray Kurzweil, *The Age of Spiritual Machines: When Computers Exceed Human Intelligence* (London: Penguin, 2000 [1999]).

[65] For the counter argument to Kurzweil, see David Moschella, "The Pace of Technology Change Is Not Accelerating," *The Leading Edge Forum,* September 2, 2015, and also D. Moschella, "The Pace of Digital Disruption Varies Widely by Industry, *ComputerWeekly.com,* September 2015.

[66] For a general introduction to Moore's Law, see The Editors of EB, "Moore's Law" and related articles, *Encyclopædia Britannica* (online), July 21, 2016, and sources presented in the *Wikipedia* article of this name. See also Robert Hallberg, "The Promise of Accelerating Growth in Technology," *Seeking Alpha*, March 23, 2012.

[67] As the *Economist* notes, in 1971 the fastest car, the Ferrari Daytona, was capable of going 178 mph, and the tallest buildings, the World Trade Center's two towers, were 1,362 feet tall. If automobile design and tall building construction had advanced, matching the pace of the growth of computing power today, cars would be traveling "a tenth of the speed of light... [and] the tallest building would reach half way to the Moon." See Economist Staff, "After Moore's Law: The Future of Computing," *The Economist* (print and online), May 13, 2010.

[68] There are literally hundreds of movies that hint at what might happen if technology runs amok, beginning with *Metropolis* in 1927, continuing through classic earlier films like *2001: A Space Odyssey, Westworld,* and *WarGames,* through the many commercial franchises (think trilogies), to depictions in this new millennium (e.g., *AI, Moon, Her, Ex Machina*). For a professor's commentary on the trend in these, see Elizabeth Alsop, "The Future Is Almost Now," *The Atlantic,* May 15, 2016.

[69] For general "nano" information, see **a)** S. Tom Picraux, "Nanotechnology," *Encyclopædia Britannica* (online), July 21, 2016, and **b)** NNCO Staff, "Nanotechnology and You," *Nano.gov,* July 22, 2016 (Arlington, VA: National

Nanotechnology Coordination Office [NNCO], United States National Nanotechnology Initiative). For an anti-cancer example, still at very preliminary stages, see Lynn Yarris, "Nanocarriers May Carry New Hope for Brain Cancer Therapy," *Lawrence Berkeley National Laboratory News Center,* November 19, 2015.

[70] Nanotech seems to desire to miniaturize almost everything. As with the other five categories, it has society-changing ethical and legal implications–think of everything evil (pathogens, bombs, etc.), except all man-made and nano (very tiny). For the risks and implications, see material presented at NNCO Staff, "Nanotechnology and You," in last note.

[71] Of course, while some celebrate the changes gene-editing may facilitate, others question our ability to constrain the technology to noble ends, and fear Pandora's Box will be forever opened. See Heidi Ledford, "CRISPR, the disruptor" (news feature), *Nature* 552 (7554), June 3, 2015, pp. 20-24. The article leads with the teaser, "A powerful gene-editing technology is the biggest game changer to hit biology since PCR. But with its huge potential come pressing concerns." For entrée into concerns regarding other new biologies, see for instance, Ian Sample, Nicola Davis, Paul Freemont, and Filippa Lentzos, "What are the risks of DIY synthetic biology? [Science Weekly podcast]," *The Guardian,* January 12, 2015, and Emily Singer, "The Dangers of Synthetic Biology," *MIT Technology Review,* May 30, 2006 (teaser, "Nobel Prize winner David Baltimore explains why building smallpox from scratch is a key safety concern in synthetic biology.").

[72] For a general introduction into additive manufacturing and 3D printing, see **a)** The Editors of EB, "3D printing," *Encyclopædia Britannica* (online), July 21, 2016; **b)** Richard D'Aveni, "The 3-D Printing Revolution," *Harvard Business Review* (print and online), May 2015; and **c)** Economist Staff, "A Third Industrial Revolution," *The Economist* (print and online), April 21, 2012. For the construction examples, see Michelle Starr, "Dubai unveils world's first 3D-printed office building," *CNET.com,* May 25, 2016, and Kendall Jones, "Here Comes the World's First Fully Functional 3D Printed Excavator," *Construction.com,* May 27, 2016.

[73] For starting points on virtual reality, see **a)** Joel Stein, "Why Virtual Reality Is About to Change the World," *Time.com,* August 6, 2015; **b)** Henry E. Lowood, "Virtual reality (VR)" *Encyclopædia Britannica* (online), July 21, 2016; **c)** Maria Konnikova, "Virtual Reality Gets Real: The Promises—and Pitfalls—of the Emerging Technology," *The Atlantic,* October 2015; and **d)** CNET Staff, "Virtual Reality 101," *CNET.com,* 22 July 2016.

[74] Among those currently investing billions are: Facebook (Oculus Rift), Google (Magic Leap), Microsoft (HoloLens) and Sony.

[75] For a brief on "big data" and "data analytics" and the scope of applications, see a) Andrew McAfee and Erik Brynjolfsson, "Big Data: The Management Revolution," *Harvard Business Review* (print and online), October 2012; b) Bernard Marr, "The Awesome Ways Big Data Is Used Today To Change Our World," *LinkedIn Pulse,* November 13, 2013 (his book is *Big Data: Using Smart Big Data, Analytics and Metrics to Make Better Decisions and Improve Performance* [New York: John Wiley, 2015]); and c) James Manyika, Michael Chui, Brad Brown,

Jacques Bughin, Richard Dobbs, Charles Roxburgh, and Angela Hung Byers, "Big Data: The Next Frontier for Innovation, Competition, and Productivity" (New York: McKinsey Global Institute, May 2011), p. 1ff.

[76] Another point about technology worth making is this: People respond to technological change differently. Some people love the "bleeding edge" so much that they devote their energies and capital to create it, or at least will camp on the sidewalk for weeks in order to be the first to have the latest gadget (e.g., tablet, game, iPhone, etc.). These are the Innovators and the Early Adopters on the diagram below. Others wait until the "bugs have been worked out," but sign up early on. Most wait longer, and some only embrace change when every other option has been taken away.

DIFFUSION OF INNOVATIONS CURVE

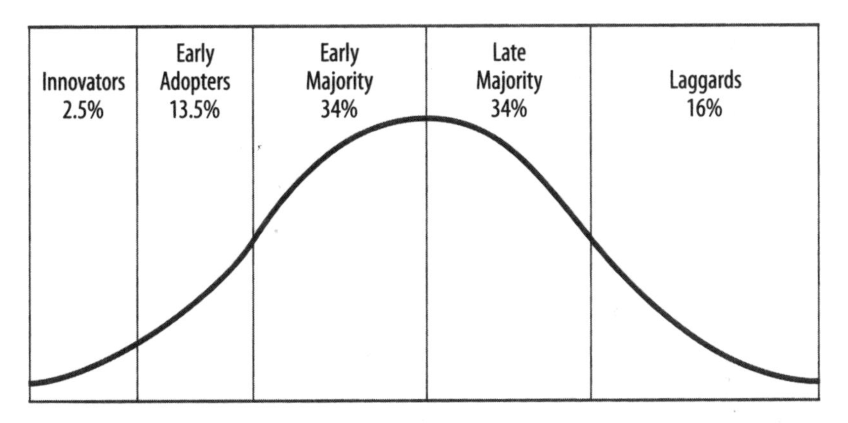

Figure: Breakdown of technology adopters, according to the classic model of Rogers. Original edition, Everett M. Rogers, "Diffusion of innovations" (New York, Free Press of Glencoe, 1962), latest, 5th Edn. (New York: Simon and Schuster, 2003). See also the sources cited at the *Wikipedia* article on diffusion of innovations, and for modern interest, Staff of the UT , "Diffusion of Innovations Theory" (Enschede, NLD: University of Twente [UT] Communication Studies, July 21, 2016).

My thoughts on this. Historically, two groups of people are famous for their resistance to technology. ***The first group*** are the Luddites, nineteenth-century British textile workers who smashed textile-making machines at the onset of the Industrial Revolution. They rebelled because they feared that the machines would put them out of work (which is essentially what happened). Today the term Luddite is used to describe (and demean) those who are slow to embrace technical advances. ***The second group*** of thoughtful "laggards" are members of the Amish faith, who prefer to wait until others uncover the unintended consequences of a new form of technology before they decide whether to embrace it or not. At the moment, relatively few others of us have developed philosophies of technology, but as the pace of technical change keeps accelerating, more people will sense the need to adopt one. Managing new technology is becoming a survival skill.

Chapter Five, *The Acceleration of Technology – Part II*

[77] In case there is someone reading who has not encountered Siri, a type of technology called a virtual assistant (or "chat-bot"), it is an AI-driven part of Apple's mobile operating systems that drives all iPhones and iPads. It is designed to use voice commands to help users access information on the web, make calls, send texts and emails, and make calendar appointments (and much more), simply by speaking to the phone. Regarding the main points of the paragraph, see Kris Hammond, "Artificial Intelligence Today and Tomorrow," *Computerworld,* April 10, 2015.

[78] Artificial intelligence is not a synthetic brain floating in a case of blue liquid somewhere. It is a set of algorithm—math equations—in a program that tells a computer and its auxiliary equipment what output to provide, or other function to perform. What separates AI from standard programs can be illustrated by comparing Deep Blue (the IBM machine that beat chess masters) with DeepMind (the Google machine that just beat the reigning Go champion). Deep Blue was programmed—that is, it was told all possible moves to make in all ppssible situations. This is how robots have functioned to date. Someone programs them to move an arm six inches to the left, grab the module, twist to the right, insert module into PC board. Repeat three hundred times each hour. These machines are not much brighter than modern toasters.

AI machines are different. A few years ago, DeepMind taught a computer how to play Atari video-games without programming it. All the computer was told was that the goal was to get a high score. Using the iterative self-modification that is central to AI, the computer messed around with the game, learned how it worked and within a few hours was able to play it better than any human. In other words, the AI machine mastered a complex task by itself, based on an initial program it was given, through modifications to that program. See Jeff Goodell, "Inside the Artificial Intelligence Revolution," *Rolling Stone,* February 29, 2016.

[79] It's worth noting a few things about AlphaGo's win. ***First,*** Google did not expect AlphaGo to be able to best human Go champions for another ten years. ***Second,*** it won with a series of moves that were described as creatively stunning by GO experts. ***Third,*** the win was unexpected in part because the game is so complicated. With about 250 possible moves each turn and a "game depth" of about 150 moves, there are approx. 250150 (10360) possible Go move variations. Even if reduced to account for obviously legal or illogical moves, the number of variations exceed the estimated number of atoms in the universe. This makes it impossible for a computer to address the problem of how to play a human and win—which AlphaGo did, for the first time, in 2016—by traditional "brute-force" (try them all) computational approaches. For entrée into this area, see **a)** Cade Metz, "Google's Go Victory Is Just A Glimpse Of How Powerful AI Will Be," *Wired,* January 29, 2016; **b)** Leon Lei, "Go and Mathematics," *AGFgo.org* [The American Go Foundation], July 21, 2016; **c)** Economist Staff, "The Economist Explains: Why Artificial Intelligence is Enjoying a Renaissance," *The Economist* (print and online), July 15, 2016; and **d)** Steven Borowiec, "Google's AlphaGo: AI defeats Human in First Game of Go Contest," *The Guardian,* March 9 2016.

[80] AI labs look like nursery schools with teams of PhDs trying to teach various "bots" how to put together a puzzle or distinguish between a piece of art and a dish towel.

[81] Jeff Goodell, "Inside the Artificial Intelligence Revolution," p. 16.

[82] Gardner, the father of this field, currently lists the math-verbal (linguistic, logico-mathematical), as well as musical, spatial, bodily kinesthetic, and interpersonal and intrapersonal, to which he has added naturalistic to an original seven. He is evaluating existential (philosophical) and pedagogic (teaching, e.g.,"As young as 2 or 3, kids already know how to teach...") as possible future additions to these eight, which have their origin in his 1983 book. The term "emotional intelligence" (EI)—which here relationship and self-awareness intelligences—a term first used in the 1960s, was formalized in models by Stanley Greenspan, Peter Salovey, and others in 1989, to elevate some parts of Gardner. The most common understandings and popular interest in EI dates to the controversial popular Bantam book by Daniel Goleman (EI as the main title), which also captures parts of the Gardner and others, applying them very broadly. See Howard Gardner, "Intelligence Isn't Black-and-White: There Are 8 Different Kinds," *Big Think* (online), January 13, 2016, which updates to his 1983 *Frames of mind: The Theory of Multiple Intelligences* (New York: Basic Books). See also sources at the *Wikipedia* pages for the persons and concepts mentioned.

[83] See notes 14 and 15 for an introduction to Mike Kelly, and the context of this extended conversation.

[84] The idea of this "singularity" traces back to a statement John Von Neumann made in the 1950s. Given the accelerating progress of technology, he argued that there was "the appearance of approaching some essential singularity in the history of the race beyond which human affairs, as we know them, could not continue." In the 1960s, I. J. Good wrote of an "intelligence explosion," resulting from intelligent machines designing their next generation without human intervention. Others followed. Today Ray Kurzweil argues that this day will soon arrive.

[85] See for instance, Heather Zeiger, "From Me to Eternity," *Salvo*, Summer 2016, p. 27f.

[86] Please note, most AI machines will not be designed to look human. There is no need for that. And machines will be programmed to feel pain so they are alerted to the fact that something is wrong.

[87] See notes 14 and 15, chapter one.

[88] The compositions of modern AI-capable devices—even just considering their integrated circuits—are of course more complex than this. The actual compositions are a veritable smorgasbord of the periodic table, having gallium, phosphorus and arsenic, among many others (depending where in the devices you look).

[89] One of those arguing most forcefully for the position I am advocating is David Gelernter, who is a pioneer

in Artificial Intelligence. For more on his views and insights see, D. Gelernter "Encounters with the ArchGenius," *Time,* March 7, 2016. See also William Carroll, "Mind the Gap: Neuroscience, Transhumanism, and Human Nature," *Public Discourse* (Princeton, NJ: The Witherspoon Institute), October 2015.

[90] Brian Fung, "Everything You Know About AI is Wrong," *The Washington Post,* June 2, 2016.

[91] Technology has replaced jobs, which over time makes life easier. In 1900, nearly forty percent of the work force was involved in growing food. Today it is down to three percent—and that three percent not only feeds the entire nation, it produces enough to allow us to ship millions of tons of food to other countries. Similarly, in1980, 400,000 people worked in the steel industry. Today 150,000 are making more steel — and doing so faster and with fewer injuries—than the group in the past. This allows people who otherwise would have to work growing food (or making steel) to make other things. But the transition is often very brutal. Charles Dickens wrote about the transition during the Industrial Revolution. It is painful to read about the suffering many endured during that time. Kelly contends that 50 percent of those who drive for a living (think: taxis, busses, semis, delivery vehicles, etc.) will be replaced by AI drivers within twenty years, and that 100 percent of these drivers will be replaced within thirty years. For Kelly, see notes 14 and 15, chapter one.

[92] Staff of Robotenomics., "Study indicates Robots could replace 80% of Jobs," *Robotenomics.com,* April 16, 2014.

[93] See Alan Mendoza, Richard Susskind, and Daniel Susskind, "Event Transcript: 'The Future of the Professions: How Technology will Transform the Work of Human Experts,'" *HenryJacksonSociety.org,* March 3, 2016. The book referred to is Richard Susskind and Daniel Susskind, *The Future of the Professions: How Technology Will Transform the Work of Human Experts* (Oxford, Oxford University Press, 2015).

[94] Jim Clifton, the Chairman of Gallup and author of *The Coming Jobs War,* argues that soon there will only be 1.2 billion jobs on the planet. Clifton contends that the job shortage will lead to a growing division between the rich and the rest and likely eventuate in a war. As I write this, the Republican primary has just concluded months of unexpected turns. I believe it is too soon to ascertain exactly what happened and why, but I am pretty sure that one of the main reasons for Donald Trump's insurgency is his promise to bring back good jobs. I also think that though much of the blame for job loss is being ascribed to immigration, that globalization and accelerating technology are bigger factors. See Jim Clifton, *The Coming Jobs War* (Washington, DC: Gallup Press, 2011).

[95] Those who argue that AI will create as many jobs as it displaces note that if you told a nineteenth century farmer that machines would make it possible for one percent of the population to grow all the food we need, they would have wondered what everyone else was going to do. They had no idea that there would be jobs for tax accountants, web designers, and public relations experts.

[96] The Editors of the WSJ, "The Future of Everything," (special issue printed, and online), *The Wall Street Journal,* December 10, 2015 and ongoing.

[97] See for instance, Staff at The Guardian, "How Likely Are you to Live to 100? Get the Full Data," *The Guardian*, July 21 2016.

[98] For a chillingly entitled description of this area, see "Emerging Technology from the arXiv," (author) "Why Self-Driving Cars Must Be Programmed to Kill," *Technology Review,* October 22, 2015. For its relationship to the classical trolley problem, see Joel Achenbach, "Innovations: Driverless Cars are Colliding With the Creepy Trolley Problem, *The Washington Post,* December 29, 2015.

[99] For Uber, AirBnB, and other businesses, see Hamish McRae, "Facebook, Airbnb, Uber, and the Unstoppable Rise of the Content Non-Generators, *The Independent* (UK), May 5, 2015. For Kodak, see Aagam Shah, "In 1998, Kodak Had...," *BesttheNews.com,* April 28, 2016, and Udo Gollup, "Graduates, Think Fast Because Change is Coming Fast," *The Virgin Islands Daily News,* May 24, 2015.

[100] See, for instance, Timothy B. Lee, "3 Threats to Incumbent Car Companies are Converging into a Tidal Wave of Disruption," *Vox.com,* May 27, 2016, and Federico Guerrini, "No Need For Insurance: How Self-Driving Cars Will Disrupt A $200 Billion Industry, *Forbes* (online), June 11, 2015.

[101] Note, Israel has not admitted to possessing nuclear weapons, but it is widely believed to have nuclear warheads. See Glenn Kessler, "Fact Checker: Iran's claim that Israel has 400 nuclear weapons," *The Washington Post,"* May 1, 2015.

[102] To the extent that there is good news on this front, several things stand out: **a)** At the moment, massive destructive power requires significant technological support. The movies may feature an evil genius hunkered down in a high-tech cave plotting to take over the world—and doing all of this alone. But that cannot be done. Complex technological systems require significant amounts of technology expertise and support. Hundreds of people must be involved. **b)** Because the military power of the United States significantly eclipses everyone else, it is unlikely that it will be subject to a direct attack by another nation. Additionally, it is likely that the conventional warfare of the future will involve so much technology (e.g., robotic soldiers, drones and cyber attacks) that it will be unlike any "conventional" war from the past.

[103] We can expect a series of pendulum-like pushes and pullbacks, as the government takes steps forward and then defends (sometimes successfully and sometimes not) the need for further domestic surveillance. See **a)** David Barrett, "One Surveillance Camera for Every 11 People in Britain" *The Telegraph,* July 10, 2013, and **b)** Henry Austin, "North Dakota becomes first U.S. state to legalize use of armed drones by police," *The Independent,* (UK) September 8, 2015. For the Dallas incident, see Sara Sidner and Mallory Simon, "How Robot, Explosives Took Out Dallas Sniper in Unprecedented Way, *CNN.com,* July 12, 2016.

[104] After reading tech magazines, visiting tech laboratories, and watching more science fiction movies in the last year than in the last twenty combined, I can say that I am more of a fan of technology than I was before I started my research. But I still have a lot of concerns. I would feel more confident offering a prediction on where we are headed if I could answer these four questions.

Question one: Will the tail wag the dog? Many let technology lead. They do X, not because X is important or necessary, but because X is new and possible. They also do X without any thought as to where X may lead, or what kind of person X will turn them into. Several years ago a young woman approached one of the staff members at our church asking for help with her husband, who was playing six to seven hours of video games every day (and consequently ignoring her). At the time we were unfamiliar with the idea that video stimulation may be a bona fide addiction, so we simply appealed to his reason. We discussed appropriate boundaries, the value of marriage, and the responsibilities of adulthood. And we lost. He told us (and her) that he would rather play video games than be married and he walked out. This is an extreme example, but it points out that technology leads some astray. I am particularly concerned about where technology leads us in view of the fact that the two groups who are consistently among the first to exploit its breakthroughs are those who build weapons and those who promote pornography.

Along these lines, Neal Postman's 1985 critique of Western Culture, *Amusing Ourselves to Death,* notes that the dark description of the future George Orwell develops in *1984* have not proven true, but those suggested in Aldous Huxley's *Brave New World* have. It's in *1984,* you might remember, that Big Brother has taken over. The future is bleak and gray because everyone is oppressed by an all invasive government that controls everything, rewrites history to stay in power, spies on its citizens twenty-four hours a day and even arrests them for committing "thought-crimes." There are those who might argue that Orwell's description of life is coming true in this country. But Postman argues that Aldous Huxley's *Brave New World* is a much closer description of what is unfolding. Postman noted that Huxley, in *Brave New World Revisited,* observed that "the civil libertarians and rationalists who are ever on the alert to oppose tyranny 'failed to take into account man's almost infinite appetite for distractions.' In *1984,* Huxley added, people are controlled by inflicting pain. In *Brave New World,* they are controlled by inflicting pleasure. In short, Orwell feared that what we hate will ruin us. Huxley feared that what we love will ruin us." It surely appears that Huxley had a clearer view of the future than Orwell. Along these lines, I am aware that some of those I know who work in high-tech fields, including some who run high-tech companies, not only limit the amount of tech they use but also tend to keep most of it out of the hands of their children. See Neal Postman, *Amusing Ourselves to Death: Public Discourse in the Age of Show Business* (London: Penguin, 2005 [1985]).

Question two: What will we unwittingly set in motion? In the April 2000 issue of *Wired* magazine, Bill Joy, then Chief Scientist at Sun Microsystems, published an article entitled, "Why the Future Doesn't Need Us." In it he warned that robotics, genetic engineering, AI, and nanotechnology could quickly get out of hand and prove more dangerous than nuclear weapons. Over the last fifteen years, Joy's article has been widely discussed and debated. Today at least a few leading thinkers express concern that someone somewhere is going to lose control of an experiment and unleash a set of problems we are not prepared to address. Stephen Hawking believes that AI "could spell the end of the human race." (Jason Pontin, "Why We Can't Solve Big Problems," p. 75f). Elon Musk, the founder of PayPal, Tesla, Space-X, and several other technology-based initiatives, refers to research in AI as "summoning the demon." (Musk recently tweeted, "Hope we're not just the biological boot loader for digital super intelligence. Unfortunately, that is increasingly probable.") And Bill Gates has declared that he shares Musk's concerns and adds, "I do not understand why some people are not concerned." See Edward Geist, "Is Artificial Intelligence Really an Existential Threat to Humanity?," *The Bulletin of Atomic Scientists,* February 10, 2016, and Michael Sainato, "Stephen Hawking. . . ," in note 11 in chapter one.

Questions three and four: What kind of havoc will bad actors initiate? As noted in the chapter, technology is placing greater power in the hands of smaller numbers of people. We seem to be moving towards a day when it's not just rogue states but also terrorist cells—and evil corporations?—that have massive destructive power at their disposal. ***Are we ready for the downsides of the upsides?*** Though advances in technology often make things better, they can be very disruptive as society transitions over to them. There will be winners but there will be losers as well. Can we keep everyone employed? What will technology do to our souls?

Chapter Six, *Changing Social, Sexual, and Marital Dynamic*

105 Because I do not like the company those titles put me with—and because I'd rather avoid the political overtones linked with that kind of language—I am being more circumspect. But I do think this issue is a very serious issue.

106 Marriage in other parts of the world tends to be quite different: arranged marriages are still common, as are large extended families. Many look to Europe for indications of the path the United States is on because of the similarities culturally. But most of Europe is less Christian, less diverse, and operates with a larger social welfare network. Let me note, I am focusing here mostly on the social, sexual, and marital dynamics of the US.

107 These discussions are especially hurtful when people who already feel like victims also feel judged.

108 Those governments that attempt to employ people to care for what they perceive as being the state's children not only go broke trying to make their system work, they cannot provide the level of love and care required for children to thrive.

109 This quote—a part of popular culture since its use as a line in CBS' *Blue Bloods*— is attributed to Michael Novak. The point is, although social scientists avoid suggesting a particular form of marriage or family as best, they agree that everyone wins when families thrive. See for instance, Michael Novak, in "Crumbling Foundations: Why the Family Unit Is Crucial to Civilization" (Crisis Magazine, December 1, 2006). Why? For starters, no one loves like Mom and Dad. Parents (biological, adoptive, and foster) sacrifice for their children in

ways others do not. And men do better in marriage—married men live longer, report less stress, earn more money, and act in ways more beneficial to society (e.g., see Staff of HMHW, "Marriage and Men's Health," *Harvard Men's Health Watch* [online, at *Health.Harvard.edu*], July 28, 2016 [2010]). It is also generally true that women in happier marriages fare better. Aja Gabel writes,

> [R]esults suggest that marriage often does make people happy and happy people are more likely to marry. [She quotes UVA psychology professor Robert Emery:] "We know that a particularly happy marriage is associated with all sorts of psychological benefits: you are less depressed, less anxious, less likely to be in trouble with the law, less likely to be engaged in drinking or drug use, and you live longer… evidence [is] that marriage is both a cause and an effect of happiness."

See Aja Gabel, "The Marriage Crisis: How Marriage Has Changed in the Last Fifty Years and Why It Continues to Decline," *The Virginia Magazine*, 2016. In today's climate it's challenging to say much about women and work without offending someone or exacerbating guilt. Women are often torn no matter what they do. If they stay at home they feel devalued; if they remain in the workforce after having children, they feel guilty about the time they are not spending with their children. That said, studies show that mothers of infants generally value the option of staying home to devote themselves to the care and nurture of their young children, a situation most common in marriages where a father provides financial support.

As noted, social science is coalescing around the idea that marriage has benefits. Ruth Graham develops this point, quoting Andrew Cherlin, a sociologist and public policy professor at Johns Hopkins:

> "It's true that there's a line some liberal sociologists won't cross… of accepting marriage as the best arrangement… [But] a growing number of sociologists… concede that in the world… today, marriage seems like the best way to give kids a stable family life." The new wave of pro-marriage scholarship is challenging orthodoxy in academic fields with reputations … of being politically liberal, and perhaps even anti-marriage… Part of the shift is because marriage itself has changed… [She continues, quoting Philip Cohen, University of Maryland sociologist and a critic of pro-marriage studies.] "Criticism of marriage as a social institution comes from the… basically compulsory system of marriage in the 1950s… When people got married who did not want to… when women's rights within marriage were much more limited, employment opportunities much less, domestic violence taken much less seriously, when rape wasn't even a crime within marriage—that system deservedly had a bad rap." The new champions of marriage disagree on how, and even whether, to encourage marriage through public policy. Nonetheless, there is an emerging consensus around an idea that would have sounded retrograde just a few decades ago: that having married parents is best for children's well-being, that marriage is beneficial for parents' psychological and economic stability, and that it should be a priority in public policy.

See Ruth Graham, "They Do: The Scholarly About-Face on Marriage," *The Boston Globe*, April 26, 2015. For Novak's quote in the text, see William Bennett, "Stronger Families, Stronger Societies," *The New York Times*, April

24, 2012 [widely cited], and Gary Bauer, "Remarks on Welfare Policy," In *Welfare: A Documentary History of U.S. Policy and Politics*, G. Mink and R. Solinger, Eds.(NYU Press, 2003), p. 510 [verbatim source].

[110] There are lots of statistics showing that children growing up without fathers active in their lives are more likely to have behavioral problems, run away from home, or become teenage parents themselves, etc. Writing in *Public Discourse*, Nathaniel Peters argues that, "Civil society… does not exist to serve the state; on the contrary, Novak argues, the state exists to serve it. Furthermore, the family is not only a place where moral capital is accrued, but also where financial capital begins. Many get their first jobs from parents, uncles and aunts, and members of their churches. Those who are serious about helping the poor need to take account of the moral ecology required for human flourishing and the structures that maintain it. See N. Peters, "Catholicism, Capitalism, and *Caritas*: The Continuing Legacy of Michael Novak," *Public Discourse* (Princeton, NJ: The Witherspoon Institute), June 2, 2015. See also, Ryan T. Anderson, "Marriage Matters," *Public Discourse* (Princeton, NJ: The Witherspoon Institute), January 15, 2014.

[111] Michael Novak argues that healthy societies depend on families to instill habits of "social trust, personal responsibility, hard work, compassion, and social cooperation" in children. And when the family fails, the government is overwhelmed. Novak paraphrases James Madison's warning, saying, "A citizenry that cannot govern its personal behavior in its private life can hardly be expected to be successful in self-government in its public life." See Michael Novak, "Crumbling Foundations: Why the Family Unit Is Crucial to Civilization," *Crisis Magazine*, December 1, 2006.

Mary Eberstadt adds: "The empirical record today on sex ubiquitously reveals the benefits of marriage and monogamy, beginning with the married partners themselves. As the sociologist W. Bradford Wilcox has shown, for example, monogamous married people score better on all kinds of measures of well-being. They tend to be happier than others. Women whose husbands are the breadwinners also tend to be happier than others, and men who are married earn more and work harder than men who are not. Conversely (as Wilcox's research has also shown), promiscuity on campus appears closely related to educational failure and other problems such as alcohol and drug consumption. Wilcox and the author Maggie Gallagher have also shown that widespread divorce and unwed motherhood ['two offspring of the sexual revolution'] are not only bad for many people but also costly for society." See Mary Eberstadt, "The Will to Disbelieve," *First Things*, February 2009.

Finally, the American Heritage's group studying Poverty, Society and Culture contends that several factors demonstrate the link between thriving families and the economic welfare of the state. See W. Bradford Wilcox, Robert I. Lerman, and Joseph Price, "Strong Families, Prosperous States: Do Healthy Families Affect the Wealth of States?" *AEI.org* (Washington, DC: American Enterprise Institute and the Institute for Family Studies, October 19, 2015).

[112] Much of what has happened recently was predicted. In a 1981 article for *Family Weekly*, futurists Alvin and Heidi Toffler claimed that the family was about to diversify. Without much effort I can point to a dozen different

"family forms" from among my friends and neighbors: Mom, Dad and children; Mom and Dad before and after children; husband and wife who do not plan to have children; single parent families, blended families, gay and lesbian couples with and without children, cohabiting couples with and without children, foster families, grandparents raising children, two single parents living together and raising children, etc. See Alvin Toffler and Heidi Toffler, "The Changing American Family," *Family Weekly,* March 21, 1981.

[113] As R.A. Mohler, states, "Western society is currently experiencing what can only be described as a moral revolution. Our society's moral code and collective ethical evaluation on a particular issue has undergone not small adjustments but a complete reversal. That which was once condemned is now celebrated, and the refusal to celebrate is now condemned. What makes the current moral and sexual revolution so different from previous moral revolutions is that it is taking place at an utterly unprecedented velocity." See R. Albert Mohler, "Essays & Perspectives: Biblical Theology and Sexuality Crisis, " *CBMW,* May 5, 2015.

[114] For starting points on this subject, see **a)** Aja Gabel, "The Marriage Crisis," cited in note 109 above; **b)** Nancy Cohen, "How the Sexual Revolution Changed America Forever," *Alternet,* February, 2012; and **c)** Ryan T. Anderson, "Marriage: What It Is, Why It Matters, and the Consequences of Redefining It," *Heritage.org* [Backgrounder #2775 on Family and Marriage], March 11, 2013.

[115] Ryan T. Anderson argues for this view in an article he wrote for the Heritage Foundation. "At its most basic level, marriage is about attaching a man and a woman to each other as husband and wife to be father and mother to any children their sexual union produces. When a baby is born, there is always a mother nearby: That is a fact of reproductive biology. The question is whether a father will be involved in the life of that child and, if so, for how long. Marriage increases the odds that a man will be committed to both the children that he helps create and to the woman with whom he does so." (Ryan T. Anderson, "Marriage: What It Is, Why It Matters, and the Consequences of Redefining It," *The Heritage Foundation — Backgrounder,* March 11, 2013). In an Oxford Press publication, Maggie Gallagher makes a similar point. She writes: "The critical public or 'civil' task of marriage is to regulate sexual relationships between men and women in order to reduce the likelihood that children (and their mothers, and society) will face the burdens of fatherlessness, and increase the likelihood that there will be a next generation that will be raised by their mothers and fathers in one family, where both parents are committed to each other and to their children." (John Corvino and Maggie Gallagher, *Debating Same Sex Marriage* [Oxford, U.K.; Oxford University Press, 2012), p. 94]). Finally, as the late sociologist James Q. Wilson wrote, "Marriage is a socially arranged solution for the problem of getting people to stay together and care for children that the mere desire for children, and the sex that makes children possible, does not solve." Marriage is society's least restrictive means of ensuring the well-being of children. Marital breakdown weakens civil society and limited government. See James Q. Wilson, *The Marriage Problem* (New York: HarperCollins, 2002), p. 41.

[116] See Elliot Pearce, "Marriage Then and Now," *The Observer,* April 14, 2013. Note also that the expansion of consumer products has made single life much easier.

[117] In *The Atlantic Monthly,* Richard Reeves wrote: "Sex before marriage is the new norm. The average American woman now has a decade of sexual activity before her first marriage at the age of 27. The availability of contraception, abortion, and divorce has permanently altered the relationship between sex and marriage." See Richard Reeves, "How to Save Marriage in America," *The Atlantic Monthly,* Feb. 13, 2014 As Stephanie Coontz, the author of *Marriage, A History and The Way We Never Were* (New York: Basic Books, 1993) puts it, "Marriage no longer organizes the transition into regular sexual activity in the way it used to.

[118] Today the average woman has fewer children and will likely live for several decades after the last one has left home. This is one of the reasons many women pursue education, which means they will marry later. It also increases the likelihood that she will be sexually active before marrying, which increases the likelihood of divorce later on—which in turn drives up the number of single people.

[119] From the "Free Love" movement of the 1860s to the Greenwich Village Bohemians of the 1910s, there has always been a subset of American society pushing for the "Free Love" expressions eventually championed by the sexual revolution. But it had been held in check by the natural consequences of sex (i.e., babies). Sex for sex's sake simply doesn't mix well with pregnancy and the care of infants. "The Pill" changed this, a point expressed a few years back on the date of its fiftieth anniversary. Numerous public intellectuals, from Walter Lippmann, Martin Marty, Francis Fukuyama, and Robert Putnam, all weighed in on its impact, noting that it's hard to think of anything else that changed life so quickly and dramatically for so many. Albert Mohler stated, "The sexual revolution was so utterly successful that most Americans living today do not even recognize that it happened." See Nancy Cohen, "How the Sexual Revolution Changed America Forever," *Alternet,* February 2012.

[120] In an interview with Marcia Segelstein, University of Texas sociology professor Mark Regnerus said, "When you take the biggest risk out of sex (pregnancy), a lot more people are going to do it. It (contraception) blew the door open to premarital sex and extramarital sex and all sorts of things. It really started with that. People don't problematize contraception like they problematize pornography. I would also say that online dating exacerbates the hookup mentality. I'm seeing its influence among forty-somethings and even fifty-somethings. I've written about the mating market and how it was split by the uptake of contraception into a pool of people interested in marriage and a pool of people interested more promptly in sex. There are more women in the former and more men in the latter. So men feel like they're in the driver's seat in the marriage corner of the mating market, and they are because they're rarer. People attribute male power to the patriarchy, but that's just not the case. I attribute male power in the mating market to the fact that contraception gives men all kinds of bargaining power that they didn't use to have." See Marcia Segelstein, "Unesteemed Colleague: An Interview with Mark Regnerus," *Salvo,* 33.

[121] On the bikini bullet: During the '50s and '60s girls were not allowed onto public beaches if they showed up in bikinis, and teen magazines in the mid-sixties maintained that "no girl of tact" would ever wear one. Public approval significantly changed following the release of the song, "Itsy Bitsy Teenie Weenie Yellow Polka-Dot

Bikini." For more information on Réard, see the citations appearing in his article and related articles at *Wikipedia*. On the Dick Van Dyke Show bullet, see, for instance, Bill Ward, "A separate sleep keeps the peace," *Star Tribune* (Minneapolis), May 19, 2012. On the Heffner obscenity charges, see Karl Klockars, "Friday Flashback: Hef's Obscenity Battle," April 10, 2009.

[122] There are a number of other reasons why marriage is so different: **a)** It is no longer expected. During the Roman Empire men were fined if they did not get married and father children because everyone needed to help grow the "tribe." One hundred years ago, men who did not marry were considered eccentric and women who did not marry were called "old maids." Today, remaining single is a legitimate lifestyle choice. **b)** Many couples now live together before marriage to see if they are compatible. Though many believe that living together is a great way to test compatibility, those who live together before marrying are more likely to divorce, if they do later marry, than those who do not live together first. **c)** One hundred years ago people did not live as long, so marriages were much shorter, and thus more likely to end with the death of a partner after ten to twenty years. And, **d)** The fact that women are not principally raising children for the bulk of their married lives means many women seek a college education, which means they are not inclined to marry until they are older. See, for instance, Cohen, "How the Sexual Revolution Changed America Forever."

[123] In an ironic twist, nudity is now so ubiquitous in our culture that *Playboy* is attempting to differentiate itself by no longer including any in its magazine.

[124] See Hanna Rosin, "Sunday Book Review: An Innocent in the Ivy League ['Sex and God at Yale,' by Nathan Harden]," *The New York Times*, August 23, 2012. The book itself is N. Harden, *Sex, God, and Yale: Porn, Political Correctness, and a Good Education Gone Bad* (New York, NY: Thomas Dunne Books/St. Martin's Press, 2013).

[125] Because it is impossible, and in some cases immoral, to control all of the variables in experiments involving human sexuality, marriage, etc., we cannot speak as definitively about cause and effect in sociology as we can in chemistry or physics. But social scientists do form hypotheses, run experiments, collect data, study correlations, and draw conclusions.

[126] Jeremy Neill writes, "The long history of human sexual restrictions changed suddenly in the West in the mid-twentieth century. In short order, the development of oral contraceptives severed the link to procreation and erased the most immediate consequence of deviant sexual activities. No longer would unwanted pregnancies be a threat to the welfare of consenting adults. At the same time, the need for offspring was erased in a dramatic stroke by the invention of government-run entitlement systems. No longer were children a necessary part of a comprehensive retirement plan. Now, for the first time ever, our retirement plans could be facilitated by other people's children. Astounding innovations in food production enabled us finally to conquer the threat of starvation, and no longer to need to produce another generation in order for our family farms to succeed. No-fault divorce offered a novel way out when couples were struggling with emotional and sexual hang-ups. And a final blow to traditional sexual norms occurred in the '60s and '70s through the rise of safe and legal abortion:

at last, it was possible to clean up the mistakes that fell through the cracks." See J. Neill, "On Human Sexuality, Conservative Victory is Inevitable," *Public Discourse* (Princeton, NJ: The Witherspoon Institute), December 2015.

[127] Richard Reeves, "How to Save Marriage in America," *The Atlantic*, Feb. 2014.

[128] In an interview in *Salvo* magazine, Professor Mark Regnerus comments on how "the mating market was split by the uptake of contraception into a pool of people interested in marriage and a pool of people interested more promptly in sex. There are more women in the former and more men in the latter. So men feel like they're in the driver's seat in the marriage corner of the mating market, and they are because they're rarer. People attribute male power to the patriarchy, but that's just not the case. I attribute male power in the mating market to the fact that contraception gives men all kinds of bargaining power that they didn't use to have." In an article in Slate titled, "Sex Is Cheap," he explored why women are willing to have sex without asking for what they often really want, commitment. There he notes that, "It doesn't feel like an individual decision to them anymore. And in some ways it's not. Because when everybody else is doing the same thing, or they perceive that everyone else is doing the same thing, they think their options are really limited. And in many real senses they are. Once upon a time, women used to ostracize other women who would be quicker to sleep with a man than they were. But that doesn't seem to be the case anymore. It signals that any idea of a cartel among women—a monopoly on sexual access—has eroded and all but disappeared. So you ask why women don't navigate the mating market in a way that is more consonant with their own interests, but it's easier said than done. They recognize that they have less power than they would like." See Marcia Segelstein, "Unesteemed Colleague," cited above, and M, Regnerus "Sex Is Cheap: Why Young Men Have the Upper Hand. . . ," *Slate*, February 25, 2011.

[129] The data in **Figure 6.1** are re-graphed from a figure provided in the article by Ana Swanson, "144 years of marriage and divorce in the United States, in one chart," *The New York Times*, June 23, 2015; see data sources noted therein.

[130] See Paul Taylor, Ed. et al., "The Decline of Marriage And Rise of New Families" [A Pew Research Center Social & Demographic Trends report], (Washington, D.C.: Pew Research Center, November 18, 2010). According to this Pew Research Center report, forty percent of people who believe that marriage is "obsolete" still hope to marry. Also note: One of the reasons that the number of married people who are married is down is because people are not marrying. Another is because a number of those who do marry later divorce. Note: The number of people divorcing is down. But there is a significant amount of confusion around this topic. There are at least two reasons. ***First,*** the "divorce rate" varies substantially depending on what you measure. There are four approaches: **a)** The Crude Divorce Rate, which refers to the number of divorces per one thousand people in a population. (The crude annual divorce rate is currently 3.6 divorces for every one thousand people in the US, regardless of age); **b)** The Percent Ever Divorced, which refers to the percentage of adults who have been divorced at any time. (Currently, 22 percent of women and 21 percent of men fit this category. Of course, some of those have remarried, so at the moment, 11 percent of women and 9 percent of men are currently divorced);

c) The Refined Divorce Rate, which is the number of divorces per one thousand married women. (In 2011, nineteen out of every one thousand marriages ended in divorce); and finally, **d)** The Cohort Measure Rate, the "40-50 percent" number that most people cite. It is not a hard, objective number, but an educated projection of the likelihood that someone marrying today will divorce at some point before they die. ***Second,*** the divorce rate has been declining in recent years in part because so many more people are living together than in the past, and though those doing so break up more often than those marrying, those breakups are not included in the number of marriages that end in divorce.

[131] The data in **Figure 6.2** are re-graphed from a graphic provided in the report by Gretchen Livingston and D'Vera Cohn, "U.S. Birth Rate Falls to a Record Low; Decline Is Greatest Among Immigrants" [A Pew Research Center Social & Demographic Trends report], (Washington, D.C.: Pew Research Center, November 29, 2012), p3.

[132] There are several reasons that the American family is smaller now than at any time in the past. Among them are these: **a)** people are marrying later (see Figure below), the fertility window is not open as long; and **b)** people who live together outside of marriage are less likely to have children.

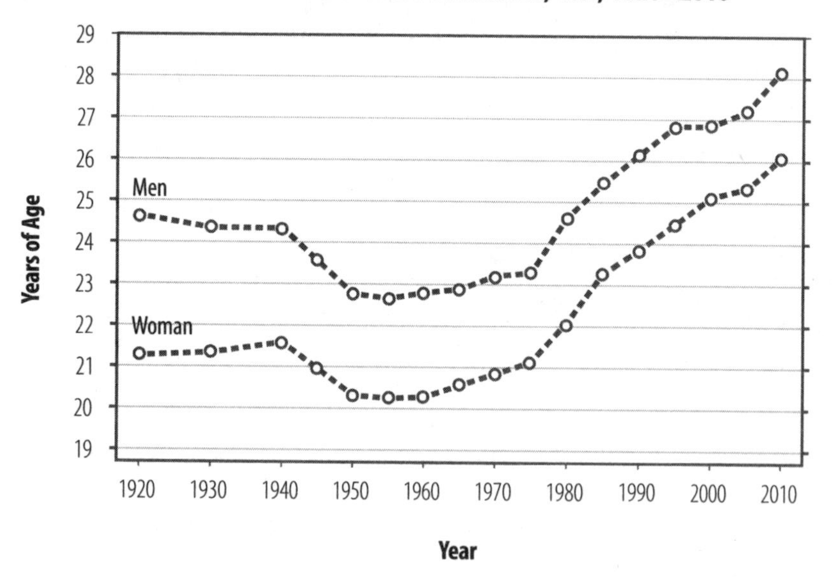

Figure. The increasing delay in the onset of marriage in the U.S. over the last fifty years, which shows a change of less than a year in the quarter century, 1951-1975, but an increase in median ages of ~4 years for men and women over the equivalent period, 1976-2000. The increases continued between 2000 and 2010 (data not shown). This 2015 analysis is by the U.S. Census Bureau, using census data. The data here are re-graphed from a figure provided by the source, Census Bureau Staff, "People and Households, Families and Living Arrangements, Marital Status," *Census.gov* (U.S. Department of Commerce, U.S. Census Bureau, October 30, 2014).

[133] According to DeParle, for births to American women under thirty, fully 50 percent are now out of wedlock. Jason DeParle and Sabrina Tavernise, "For women under thirty, most births occur outside marriage," *New York Times*, 17 Feb 2012.

[134] The data in **Figure 6.3** are re-graphed from a figure provided in the article by Jason DeParle and Sabrina Tavernise, "For Women Under 30, Most Births Occur Outside Marriage," *The New York Times*, February 17, 2012. See data sources noted therein. For further discussion see Robert Rector, "How Welfare Undermines Marriage and What to Do About It" [Issue Brief #4302 on Welfare and Welfare Spending], *Heritage.org*, November 17, 2014.

[135] See Bailey, "Is Modern Society. . .," [Section, "Some examples of decline," Subsection 13, "Out-of-wedlock births and single-parent households"].

[136] Charles Murray, *Coming Apart: The State of White America*, 1960-2010 (New York, NY: Penguin Random House/Crown-Forum, 2012). See also Nicholas Confessore, "Tramps Like Them: Charles Murray Examines the White Working Class in 'Coming Apart,'" [book review], *The New York Times*, February 10, 2012.

[137] In "They Do. . ." *Boston Globe* correspondent Ruth Graham writes: "A broad view of the statistics shows marriage in undeniable decline in America. But in fact, the institution is thriving among one group: the educated. People with at least a bachelor's degree wait longer to marry than other groups, but most of them do marry, and their divorce rates are low. Where marriage is really struggling is among those with lower education levels. Only 13 percent of high-school dropout millennials have had all their children within marriage, compared to 68 percent of millennial parents with at least four years of college experience." See Ruth Graham, "They Do: The Scholarly About-Face on Marriage," *The Boston Globe*, April 26, 2015.

[138] The following statistics are also germane. Between the '70s and the '90s, the divorce rate among the college educated fell from 15 to 11 percent. In contrast, the divorce rate among those with only a high school education, already high, rose slightly, from 36 to 37 percent. See Gabel, "The Marriage Crisis."

[139] Writing in *The Atlantic*, Richard Reeves observes: "The new upper class still does a good job of practicing some of the virtues, but it no longer preaches them. It has lost self-confidence in·the rightness of its own customs and values, and preaches non-judgmentalism instead. [They] don't want to push their way of living onto the less fortunate, for who are they to say that their way of living is really better? It works for them, but who is to say it will work for others? Who are they to say that their way of living is virtuous and others' ways are not?" Richard Reeves, "How to Save Marriage in America" *The Atlantic Monthly*, Feb. 13, 2014.

[140] Barring religious revival, sexual morality seldom becomes more traditional. It almost always drifts towards more libertine expressions.

[141] Dale Kuehne is a participant in the Q Ideas conferences and is a professor at Saint Anselm College in New Hampshire. See his web pages at either of those sites.

[142] Matt Ridley, *The Rational Optimist: How Prosperity Evolves* (New York, NY: Harper, 2011), p 13.

[143] Marriage may be declining, but nearly half of the people in the United States are married so the trend towards away from marriage could continue for a long time. Furthermore, birth rates are down but they could drop much lower.

[144] Jeremy Neill, "Our Great Sexual Adventure: Where Does It End?," *Public Discourse* (Princeton, NJ: The Witherspoon Institute), 2015. After stating he would still be buying shares, Neill writes: "But stocks don't rise forever, and neither will the sexual revolution. Eventually, its chaotic and destructive consequences will cause things to swing back in another direction."

[145] Neill, in "Our Great Sexual Adventure," argues: "History might never exactly repeat itself, but its many phases do resemble each other. The historical record overwhelmingly suggests a return to stricter sexual standards. Numerous cultures have undertaken large-scale licentious experiments before, with similarly destructive results. Late-stage Rome is today a favorite example among contemporary conservatives, but eighteenth-century England and France were also debauched, as were twentieth-century Weimar Germany, ancient Persia, Babylon, and certain parts of classical Greece." Later he adds, "When a large-scale social movement—such as the "second" sexual revolution of the last fifteen years—is so very, very different from all that has preceded it, it often is unsustainable beyond a couple of generations. Think of the French Revolution, America's alcohol prohibition experiment, National Socialism, and even our world's century-long flirtation with Communism. The forward momentum of the sexual revolution will soon encounter some major and chaotic consequences that will compel—not *convince* via discourse, mind you, but *compel*—our society to return to more normalized restrictions."

[146] *Obergefell v. Hodges* is the U.S. Supreme Court decision of 2015, decided 5–4, finding that same-sex couples have the fundamental right to marry, as guarantee derived by the court majority from the Due Process Clause and the Equal Protection Clause of the U.S. Constitution's Fourteenth Amendment. See Lyle Denniston, "Opinion analysis: Marriage now open to same-sex couples [*Obergefell v. Hodges*]," *SCOTUSblog*, June 26, 2015.

[147] There are several other things to note here: **a)** Though marriage will decline, I do not believe it will completely fade away. It will remain vibrant among religious people and to a lesser extent among college graduate; **b)** In ways similar to what is currently in place in Europe, I believe that churches (synagogues, mosques, etc.) will stop performing weddings as agents of the state and will instead perform blessings on marriages already solemnized by a clerk of the court. These blessings will look just like a wedding, however, the minister will no longer sign a wedding license. I also suspect that religious marriages will look increasingly different from other non-religious marriages.

[148] Because life-expectancy continues to go up, the world's population continues to climb and will likely do so for the next thirty years. But most demographers believe that the United States will begin experiencing population decline soon after that, and that the global population will begin to drop around 2050. In Jonathan Last's cleverly titled book, *What to Expect When No One's Expecting* (see in full below), notes that birthrates have been declining for several hundred years and in the last few decades the pace of the decline has accelerated dramatically. While India, sub-Saharan Africa, North Africa, and the Middle East continue to enjoy the benefits of demographic increase, and the United States, Latin America, and Southeast Asia are experiencing only modest deterioration in their dependency ratios, all of Europe is below the 2.1 replacement rate and seventeen European nations are at what demographers call "lowest low" fertility (1.3 or less). I do not believe that any society in human history has recovered after dropping that low. At this moment those in demographic crisis have two choices: economic decline or demographic transformation. Though some technologists argue that the merger of AI and robotics will take care of the problems associated with a declining population, I am skeptical. I believe the population decline will be painful. Finally, one other note: The United States is positioned better than other developed nations because: although our birthrate is below the 2.1 children/woman replacement level, it is higher than most other western nations; furthermore, many people still want to move to the United States. As a result, the Census Bureau projects that the population of the United States will grow by twenty percent between 2010 and 2030. However, most of this growth will be the result of people living longer and immigration. I see little to suggest that a baby boom is in our future. See J. Last, *What to Expect When No One's Expecting: America's Coming Demographic Disaster,* (New York: Encounter Books, 2014).

[149] For a starting point here, see Livingston and Cohn, "U.S. Birth Rate Falls to a Record Low; Decline Is Greatest Among Immigrants," and Ashifa Kassam, Rosie Scammell, Kate Connolly, Richard Orange, Kim Willsher, and Rebecca Ratcliffe, "Europe needs many more babies to avert a population disaster," *The Guardian,* August 22, 2015.

[150] See again, Murray, *Coming Apart,* and Confessore, "Tramps Like Them."

[151] Second president of the United States, John Adams, and many other founding fathers were wary of the expectation that they could create a government providing such stability, unless it was in the context of the moral restraint associated with a religious electorate. See the mention of John Adams again in chapter eight.

[152] The data in **Figure 6.4** are re-graphed from a figure appearing in the article by Josh Barro, "Lessons From the Decades Long Upward March of Government Spending, *Forbes,* April 16, 2012, see sources of data therein.

[153] See for instance, Sabrina Eaton, "U.S. Rep. Jim Renacci Introduces Bill to Put National Debt in Plain Sight." *Cleveland.com,* July 07, 2016.

[154] How will we fare in the future? I think the answer pivots in large part around three questions: The first

question is: ***How much longer will the sexual revolution continue?*** Kuehne (see note 141 above) argues that when debt and sexual excess overwhelm a culture it falls fast. I have no idea if he is right or not, but every sign that people are waking up to the problems sexual libertinism is presenting is a good one. At the moment there appears to be growing awareness that Internet porn is rewriting sexual norms in tragic ways (e.g., it makes males more accepting of rape; it leads to less interest in marriage; it promotes infidelity; it leads to less interest in having children, etc.). There is also a growing awareness that first exposure to pornography is happening at younger and younger ages and that children who have been exposed to pornography are: likely to become sexually active early, have more sexual partners, etc. Internet Porn is a tragedy. The growing awareness of this is helpful. The second question is: ***Will the haves help the have-nots?*** A second issue to monitor is how the wealthy interact with the under-resourced. As with the sexual revolution, there are reasons for encouragement. Corporations (and other entities) are increasingly encouraging volunteerism. And we have certainly seen a spike in service outside the walls of the church. There need to be obvious ways for the Have-Nots to climb. If those are lacking, anger and unrest grows. The third big issue to monitor is: ***How will we navigate our debt issues?*** Personal debt aside, our state and federal financial challenges—coupled with our pension liabilities—will eventually require action. The ideal way forward is to grow our way out of our debt. But a declining population makes growing the economy unlikely.

Chapter Seven, *Globalization*

[155] For Friedman's definition, see Thomas Friedman, "The Lexus and the Olive Tree" [summary of book by same name], *ThomaslFriedman.com,* July 21 2016. Other definitions for globalization range from "colonialization" to "the integration of markets, nation-states, and technologies in ways that enable individuals to reach around the world farther, faster, deeper, and cheaper than ever before" (see Shanza Khan and Adil Najam, "The Future of Globalization and its Humanitarian Impacts," *Boston University Center for the Study of the Longer-Range Future* (online materials), November 2009, p. 6). The United Nations defines globalization as "free trade, which includes the removal of tariffs and other impediments to the free flow of capital, goods, labor, and services." What they all have in common is the idea that trade and culture exchange are expanding in ways that national economies and national boundaries have less significance than before.

[156] Michael Sacks writes, "Global capitalism is a system of immense power, from which it has become increasingly difficult for nations to dissociate themselves. More effectively than armies, it has won a battle against rival systems and ideologies, among them fascism, communism, and socialism, and has emerged as the dominant option in the twenty-first century for countries seeking economic growth. Quite simply, it delivered what its alternatives merely promised: higher living standards and greater freedoms. Countries that have embraced the new economy — among them Singapore, South Korea, Taiwan, Thailand, China, Chile, the Dominican Republic, India, Mauritius, Poland, and Turkey—have seen spectacular rises in living standards." See M. Sacks, *The Dignity of Difference* (Bloomsbury Academic, 2003), p. 28.

[157] People have long been traveling to distant lands as explorers, soldiers, sailors, traders, and vagabonds. And there was a spike in globalization prior to WWI, but since WWII globalization has risen to new levels. Since 1959, world trade has increased roughly twentyfold from $320 billion to $6.8 trillion.

[158] Between 1980 and 2005, the ratio of world exports to world GDP has more than doubled, see Khan and Najam, "The Future of Globalization. . .," p. 25, cited this chapter above. Gilpin notes, "Since the end of World War II, *international trade* has greatly expanded. Whereas the volume of international commerce had grown by only 0.5 percent annually between 1913 and 1948, the value of world trade has increased from $57 billion in 1947 to $6 trillion in the 1990s. In addition to the great expansion of merchandise trade (goods), trade in services (banking, information, etc.) has significantly increased during recent decades. With this immense expansion of world trade, international competition has greatly increased. Although consumers and export sectors within individual nations benefit from increased openness, many businesses find themselves competing against foreign firms that have greatly improved their efficiency. . . . As integrative processes widen and deepen globally, some believe that markets have become, or are becoming, the most important mechanism determining both domestic and international affairs. In a highly integrated global economy, the nation-state, according to some, has become anachronistic and is in retreat." See Robert Gilpin, *The Challenge of Global Capitalism: The World Economy in the Twenty-first Century* (Princeton University Press, p. 5).

[159] ***Those opposing globalization*** contend that it exploits workers in the developing world by subjecting them to poor working conditions (i.e., "sweat shops") and paying them salaries that would not be allowed back home; they also argue that few of the profits are repatriated or invested in the communities whose labor and resources they consume, and that globalization is a blatant form of colonialization. ***Those supporting globalization*** note that since China opened their markets to the world in 1980, Chinese per capita personal income has risen from $1,420 to $4,120. See Trading Economics Staff, "China Disposable Income per Capita, 1978-2016, *TradingEconomics.com,* July 21, 2016.

[160] Marcia Segelstein, "The Apologist," *Salvo,* Vol. 7, Winter 2008.

[161] As Sacks has written, "Nor has globalization been only economic. It has been cultural as well. The Internet, cable and satellite television (and the global presence of mega-corporations) have brought about a huge internationalization of images and artifacts—what Benjamin Barber calls McWorld. The same jeans, T-shirts, trainers, soft drinks, fast-food stores, music and films can be seen on the streets of almost every major city on earth, and in many remote villages also. The world they represent is overwhelmingly American, and it tends to overwhelm local traditions, which are preserved, if at all, as tourist attractions. This too is deeply threatening to the integrity and dignity of non-Western civilizations. It is often difficult for the West to understand how alien, even decadent, its culture appears to those who experience it as a foreign importation. See Sacks, *The Dignity of Difference,* p. 30f, cited this chapter above.

[162] Some go so far as to define globalization as "the movement of capitalism across the globe." See IASC Staff, *Hedgehog Review: Annotated Bibliography of Religion and Globalization* (Charlottesville, VA: Institute for Advanced Studies in Culture, 2002) p. 116.

[163] "The significant decline in poverty over the last twenty years is one of the great unheralded triumphs of the twenty-first century. As recently as 2003, nearly 35 percent of the world's inhabitants were considered poor. By 2012, even after two stock market meltdowns, that percentage has dropped to 12. See Staff of the IBD, "Freedom = Growth," *Investor's Business Daily* [IBD], May 27, 2016.

[164] On allusions to capitalism as the worst economic system, see John B. Caron, Capitalism Worst System Except For Others, *National Catholic Reporter,* August 25, 1995, and Matt Barnes, "Capitalism: The Worst Economic System, Except For All the Others," *The Pitt News,* August 26, 2014. The Frank Capra film, *It's a Wonderful Life* is still in release, after its 60th anniversary. Its ASIN is BOOOHEWEJO.

[165] See for instance, Vincent Trivett, "25 US Mega Corporations: Where They Rank If They Were Countries," *Business Insider* (online), June 27, 2011.

[166] Capitalism is Darwinian. It shows no loyalty to person or place. It is not inclined to honor culture or care for the weak, unless honoring culture and caring for the weak is somehow good for the bottom line.

[167] Quote taken from Thomas Friedman, "Doing Our Homework," *The New York Times,* June 24, 2004. See also Thomas Friedman, *The World Is Flat: A Brief History of the Twenty-first Century* (New York: Macmillan-Farrar, Straus and Giroux, 2005).

[168] One of the criticisms of globalization is that decisions about the people (or a community, the land, etc.) are often made by people who live on the other side of the world and will not be affected by what happens. Sacks writes, "The feudal lord and the industrialist, however exploitative, had at least some interest in the welfare of those they employed. Today's global elites have little connection to the people their decisions affect. They do not live in the same country as those who produce their goods. They may have little if any contact with those who buy them, especially when purchasing is done through the Internet." Also, "Even if our moral imagination were in good order, many of the issues posed by globalization would tax the resources of conventional ethics. This was the case argued, in the 1980s, by one of the prophets of the technological age, Hans Jonas. In his *The Imperative of Responsibility,* he argued that modernity was redefining the parameters of human action and choice. Traditional ethics saw action in terms of its immediate effect on others who were usually close to hand. Nowadays we are aware of the long-term effects of human behavior on those who are distant from us, whether in space or time (generations not yet born). We also face the problem of thinking about actions, insignificant in themselves, which none the less have a cumulative impact, such as the effect of aerosols or fuel emissions on the earth's atmosphere." (Sacks, p. 32,34). Khan says something similar in his report (Khan and Najam, p. 15).

[169] Global capitalism almost always favors a low-wage, high-growth economy. In the 1880s and 1890s it was the United States. After the war it was Japan. Then China. As China begins to fade, more than a dozen countries seem poised to be next—some are betting on those around the Indian Ocean Basin, others expect Eastern Africa to emerge, still others favor Southeast Asia or Latin America.

[170] I am indebted to the chapter, "The Major Cause of Change," in Price Pritchett's booklet, *Mindshift: The Employee Handbook for Understanding the Changing World of Work* (Dallas: Pritchett, 2012), for this helpful explanation. Note, the notion of business transaction as communication dates at least to Adam Smith, see Russell Weinstein, "Adam Smith (1723-1790)," *The Internet Encyclopedia of Philosophy,* July 22, 2016.

[171] The inaugural address of 20 January 1961 is available at the John F. Kennedy Memorial Library, see JFK Memorial Library Staff, "Inaugural Address, 20 January 1961," *JFKLibrary.org,* July 21, 2016. On shopping see, Frank Pellegrini, "The Bush Speech: How to Rally a Nation," *Time,* September 21, 2001.

[172] Jay Walker-Smith, the president of the marketing firm Yankelovich, claims that the number of ads climbed to five hundred per day in the '70s and that is now approaching five thousand.

[173] Though globalization had great momentum at the start of the twentieth century, it stalled when the world plunged into war. At present, the range of options available to globalization spans from a plethora of small, isolated, uncooperative, city-states at the Balkanized end of the spectrum to a one-world system (with one language, one currency, one government, etc.) at the other. We are somewhere in the middle. This means that there is room to move towards greater cooperation. At the same time, there will always be opposition. In his Harvard Business Review Press book, *World 3.0: Global Prosperity and How to Achieve It,* (Cambridge, MA: Harvard University, 2011) Pankaj Ghemawat, the professor of strategic management and Global Strategy at IESE Business School at the University of Navarra in Barcelona, argues that though there is plenty of room for globalization to advance (Ghemawat notes that only 10 to 25 percent of economic activity is international) our tribal natures fight against further cooperation. He calls suggestions otherwise to be "globaloney." See, Michael Shermer, "Globaloney," *Scientific American,* August 2011.

[174] In the Figure below, I illustrate the opposing forces that will determine if globalization will continue. Note the significant forces that contribute to a pushback that globalization is experiencing.

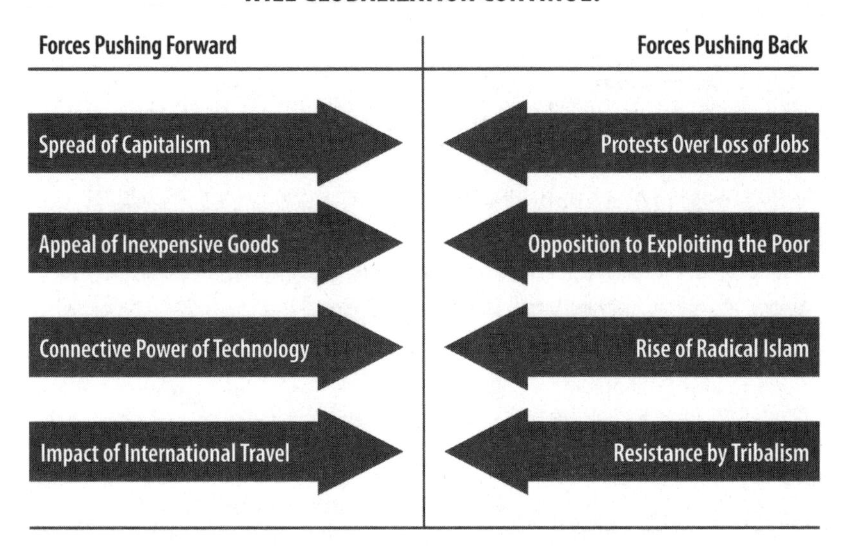

Figure. Some major component forces resisting unchecked expansion of globalism, with components from commerce and technology arrayed significantly on the left and social, cultural, and religious components arrayed on the right. For additional nuance, see the further relevant citations of this chapter, and Charles O. Lerche III, "The Conflicts of Globalization, *International Journal of Peace Studies,* Vol. 3, No. 1, January 1998.

[175] As with the previous two chapters, I believe there are a few areas to watch for indications as to how things will play out. The questions are these: ***Will the Haves help the Have-Nots?*** I asked this before, but it is important enough that I raise it again. Will those who are "winning" help those who are "losing?" What kind of safety net will they build for those who are struggling? If those who are pushed down by globalization see a path forward, they are more likely to work within the system than try to subvert it. ***As corporations grow in power, will they be good global citizens?*** Businesses are not countries (i.e., their objective is not to promote justice, provide for the common defense, or safeguard the rights of the weak). The goal is to maximize share-holder value. Many companies are funneling dollars and volunteer hours back into the community. Will they do this in a meaningful way going forward? ***In the United States, will anyone be able to govern?*** During times of transition and disruption, such as are going on now, it is not uncommon for people to move towards more extreme political views. This makes it harder for the center to hold. Alongside this country's unusual election, parties on the far left and far right are gaining rowing. Will the center hold? Will anyone be able to unite nations and lead them forward?

[176] Friedman notes that no two countries that both have a McDonalds restaurant have fought in a war against each other since opening their McDonalds. He writes about this in *The Lexus and the Olive Tree,* a 1999 book in

which he posits that there are two struggles currently defining the world: the drive for prosperity and development, symbolized by the Lexus, and the desire to retain identity and traditions, symbolized by the olive tree. In recent writings he has revised that to say that no two countries that participate in a supply chain have gone to war.

[177] See, for instance, Craig S. Smith, "Globalization Puts a Starbucks Into the Forbidden City in Beijing," *The New York Times*, November 25, 2000.

[178] We are likely to feel rushed because we are – in a world where globalization is shaping things, speed is a competitive advantage – and also because the increased number of options open to us makes us feel the

[179] Kevin Kelly, "The 24 Ways We're Tracked on a Regular Basis Reveal Something Disturbing About the Future, *TechInsider.io*, June 27, 2016..

[180] Look back to chapter one, and the original overview of the book, where we refer to Collins, as quoted by Buford, "Active Energy".

Chapter Eight, *The World of Swirling Ideologies and Beliefs*

[181] In a 1968 article in the *New York Times*, prominent sociologist Peter Berger predicted that, "By the twenty-first century, religious believers are likely to be found only in small sects, huddled together to resist a worldwide secular culture." Just over thirty years later Berger wrote, "the assumption that we live in a secularized world is false. The world today, with some exceptions. . . is as furiously religious as it ever was, and in some places more so than ever." He goes on to say, "This means that a whole body of literature by historians and social scientists loosely labeled 'secularization theory' is essentially mistaken." Hence, during the intervening years, Berger had reversed his position (as did others). See, Peter L. Berger, "A Bleak Outlook Is Seen for Religion," The New York Times, February 25, 1968, p. 3, and Berger, "The Desecularization of the World: A Global Overview," in *The Desecularization of the World: Resurgent Religion and World Politics*, P.L Berger, Ed. (Washington, DC and Grand Rapids, MI: Ethics and Public Policy Center and Eerdmans, 1999), p. 1-18, esp. 2. Note, many who originally advanced the Secularization Thesis not only felt that religion would fade, they expected that as it did, a Utopia would emerge.

[182] Fraser continues, "[T]he twentieth century was numerically the most successful century for religion since Christ was crucified and Muhammad gave his farewell sermon on Mount Arafat. By 2010, there were 2.2 billion Christians in the world and 1.6 billion Muslims, 31 percent and 23 percent of the world population respectively. The secularisation hypothesis is a European myth, a piece of myopic parochialism that shows how narrow our worldview continues to be." See G. Fraser, "The World is Getting More Religious, Because the Poor Go For God," *The Guardian*. 26 May 2016. For the philosopher's earlier misguided declarations, see Friedrich Wilhelm Nietzsche, *Die fröhliche Wissenschaft* [The Merry Science] (Chemnitz, DEU: Verlag von Ernst Schmeltzner, 1882), §107, §116.

[183] Earlier I mentioned that the CIA was surprised by the revolution that deposed the Shah of Iran and replaced it with an Islamic state (see the opening of chapter one). It turns out they ignored this fourth glacier. Recall, when the CIA was asked how it had missed the clues pointing to such a major uprising, they explained that although they had been paying close attention to the political, economic, and the military affairs in the country, they had ignored the religious ones because they did not think they mattered (seeing them as anachronistic; Torrey Froscher, "[Discussion of] Robert L. Jervis' 'Why Intelligence Fails," cited in full above). No one would make that mistake today. There is a growing awareness of how important belief is, both for individuals and, when taken collectively, for countries.

[184] Kant argued for the difference between that which our senses perceive (*phenomenon*) and that which exists in itself, independent of our senses (*noumenon*). For an entry into Kant's work in relation to religion (and connection with primary sources), see Wayne P. Pomerleau, "Immanuel Kant: Philosophy of Religion," *The Internet Encyclopedia of Philosophy,* July 22, 2016. For an approachable discussion of *noumenon* and *phenomenon,* see Jonathan Bennett, *Kant's Analytic,* (Cambridge, UK: Cambridge University Press, 1966), p. 22-27. For a biographical sketch that develops his ideas chronologically, see Michael Rohlf, "Immanuel Kant," *Stanford Encyclopedia of Philosophy,* July 22, 2016.

[185] Carl Sagan, *Cosmos* (New York: Random House-Ballantine, 2013 [1980]), p. 1, xvii.

[186] In his recent polemic against Christianity, *God Is Not Great: How Religion Poisons Everything* (McClelland & Stewart, 2008), the late author and prominent Anglo-American atheist Christopher Hitchens wrote, "Our belief is not belief" (p. 5). To which I reply, "The [man] doth protest too much, methinks" (Hamlet III.ii). Of course, to Hitchens, belief is a belief. In fact, philosophers would point out that his assertion that *his* belief is not belief is self-stultifying. That is, it is just like the comments, "My brother has no siblings," or "I always lie." It can never be true. (Regarding this brutally honest, strident polemicist now deceased, see recent written works by his brother, Peter Hitchens, and friend, Larry Taunton.)

[187] I've camped on this point because, whatever worldview you embrace, it is human nature to believe that it is correct and that you are being reasonable and objective. The truth is, we all have a worldview, and while we can build a case for the accuracy of our worldview, *no worldview can be proven.* We all accept our starting point, at least in part, by faith. An example of movement away from this awareness can be found in the evolution of perspectives regarding the series of documents called the Humanist Manifestos. In 1933 the first Humanist Manifesto was drafted by thirty relatively prominent people. It was a renunciation of traditional theism and the faith it implied, especially faith in a god or God that answered prayer. But the writers of the first Manifesto were adamant that secular humanism was a religion. In other words, it was just faith that went in a very different direction, for instance, believing the idea that man is inherently good and will, over time, evolve into a better species. Its next version, Manifesto II, equivocated on this point, and Manifesto III—which was released just prior to the new millennium—is adamant that secular humanism is not a religion.

[188] A worldview helps us assign meaning to experiences and provides a framework for making important decisions. Worldviews are not collections of facts that can be proved in a laboratory. Rather, they are a set of assumptions we accept as a matter of belief, which then shape how we see and understand everything else. Scholarly thought about worldviews has taken place in both secular and religious contexts, and with western and global emphases; another definition and source that provides further detail and emphasis is Jim Sire's, who defines worldview as "a commitment... that can be expressed as a story or in a set of presuppositions... assumptions which may be true, partially true or entirely false... which we hold (consciously or subconsciously, consistently or inconsistently) about the basic constitution of reality... that provides the foundation on which we live and move..." See James W. Sire, *The Universe Next Door: A Basic Worldview Catalog*, 5th edn., (Downwers Grove, IL: IVP Academic, 2009 [1976]), p. 20, and references therein.

[189] The example I use to illustrate our assumptions is from Jim Leffel of *Xenos*. Imagine what would happen if you were to ask a college student why he was enrolled in a particular class. The conversation might unfold like this:

You: Why are you taking philosophy?
Student: To satisfy a humanities requirement.
You: Why do you want to satisfy a humanities requirement?
Student: So I can graduate!
You: Why do you want to graduate?
Student: To get a job!
You: Why do you want a job?
Student: So I can make money.
You: Why do you want to make money?
Student: So I can buy stuff!
You: But why do you want to buy stuff?
Student: So I can be happy.
You: And why do you want to be happy?

It is at this point that the conversation stops because the student has been stunned into silence. It is inconceivable to him that anyone would not want to be happy, or even question why others would want to be happy. However, though he is unable to defend this assumption, he accepts it without being able to prove it. See J. Leffel, "Understanding Basic Beliefs," *Xenos.org*, 1994.

[190] The law of noncontradiction holds that a statement cannot be both true and not true in the same way at the same time. See the "Laws of Thought" article at *Encyclopedia Britannica* (online), and James Danaher, "The Laws of Thought," *The Philosopher*, Vol. 92, No. 1 (Spring), 2004.

[191] I tried to argue, to no avail, that college was the perfect time for students to have their ideas challenged, and to be encouraged to think about life's biggest questions.

[192] Rather than commit you to a philosophy PhD program requiring learning seven languages, I offer the following to help. It is a chart of worldviews. Note: the chart does not include a category for agnosticism; it's hard to provide definitive answers for people who do not believe in definitive answers.

WORLDVIEW CHART

	Atheism	Pantheism
Examples:	Naturalism, Existentialism, Secularism, Nihilism & Humanism	Hinduism, Buddhism, New Age Spirituality
What is real?	There is no God. There is no soul or spirit. The cosmos is all there is.	Only 1 spiritual dimension exists. All else is illusion. Spiritual reality is eternal, impersonal and unknowable. Everything is part of god and god is in everything and everyone. All of creation – people, plants, animals and rocks – are equally divine.
Who am I?	A purely physical being.	A spiritual being temporarily weighed down by a physical body. Ultimately man is impersonal and will meld into the cosmic consciousness.
Where did I come from?	Humans are the chance product of an un-directed evolutionary process.	There is no consistent answer. Many believe in some sort of theistic evolution.
What is expected of me?	There are no absolute expectations, though different subsets of atheism may have some.	Be true to yourself. Worship the creation.
What happens when I die?	Our bodies decay and we cease to exist.	There is no consistent answer. Some believe in heaven, others reincarnation.
How do I determine right and wrong?	Science provides laws that govern matter, but there is no basis for right/wrong. Values are simply preferences.	There is no consistent answer.
How can I be sure?	Truth is usually reserved for those things that can be proved empirically.	Inner truth.

Polytheism	Monotheism
Thousands of religions	Judaism, Islam, Deism, Christianity
There are many different spirit beings who govern what goes on. Gods and demons are the real reason behind "natural" events. Material things are real but have spirits associated with them.	An infinite, personal god exists. He created a finite material world. Reality is both material and spiritual. The universe as we know it had a beginning and will have an end.
The creation of the gods, just like the rest of the creatures on earth. Some tribes may have a special relationship with some gods who can protect or punish them.	An eternal, personal, spiritual and biological being who was made in the image of God.
There is no consistent answer.	Mankind was created by an infinite, personal God.
To avoid angering the gods. (It is just as important to avoid angering the evil gods as the good ones.)	In different ways all 3 of the major monotheistic faiths call on its followers to serve God and others.
There is no consistent answer.	I will face God and be judged for my life.
Truth about the natural world is discovered through a shaman who has visions telling him what the gods and demons are doing.	Moral values are the objective expression of an absolute moral being.
Truth about God is known through revelation. Truth about the material world is gained via the 5 senses in conjunction with rational thought.	Truth about God is known through revelation. Truth about the material world is gained via the 5 senses and rational thought.

[193] To dive into this subject (reader beware, this is 920 feet, center of Lake Michigan deep), see David A. Truncellito, "Epistemology," *The Internet Encyclopedia of Philosophy,* July 22, 2016, and Matthias Steup, "Epistemology," *Stanford Encyclopedia of Philosophy,* December 14, 2005.

[194] Such question sets are an interest of philosophers, anthropologists, and others, and they are generated for various reasons (including, in the case of some Jewish writings, as an early meditation on our eventual deaths). They exist in various formulations; this set is my own take, in size and composition, with some overlap with the lists of others—as there must be, when thoughtful people aim for the fundamental questions that inform our existence. For a comparison to a list that dates back to the 1970s, see the was-7-now-is-more list of Jim Sire, *The Universe Next Door,* p. 22f.

[195] People fall into one of five categories based on the way they answer ***Question 1., What is the most important thing?*** And because "most important" and "god" go hand in hand, the categories are: atheism, agnosticism, monotheism, polytheism and pantheism. ***Atheism*** is the belief that there is neither a god nor any type of supernatural activity in the universe. ***Agnosticism*** is the belief that ultimate truth is not knowable. The term is also used of those who have not yet decided what they believe—a category that is further divided between those who are looking for an answer and those who are not. ***Pantheism***—a combination of the Greek words pan, which means all, and theos, which means god—is the view that "god is all" or "all is god." It has been popular in the East for several thousand years, being expressed through Hinduism and some forms of Buddhism, and it began to gain a serious following in the West with the New Age movement (Shirley McClain in the 1980s, Oprah Winfrey in the new millenium). Today pantheism is embraced by some environmentalists and by others who embrace a general spirituality of life. ***Polytheism*** is the belief in the existence of many gods, supernatural beings that are usually understood to be both eternal and spiritual. In some cases one deity is superior to all others. Polytheism was common in the ancient world where it was embraced by both the Greek and the Romans. Today it is advocated by Hindus, some Buddhists and by many indigenous African religions. ***Monotheism*** is the belief that there is only one god, an eternal being who is separate from the created world. There are three main monotheistic religions—Judaism, Christianity, and Islam. All trace their origin back to Abraham of the Hebrew Bible (the Christian Old Testament), and all consider this book to be sacred.

[196] This statement by Lewis—late Professor of Medieval and Renaissance Literature at Magdalene College, Cambridge, and popular British Christian apologist and story writer—was given in answer to a question asked of him by a factory worker in the U.K. in the 1950s or early 1960s: "Which of the religions of the world gives to its followers the greatest happiness?" To this, Lewis replied: "While it lasts, the religion of worshipping oneself is the best." He goes on to tell this story, and explain further: "I have an elderly acquaintance. . . who has lived a life of unbroken selfishness and self-admiration from the earliest years, and is, more or less, I regret to say, one of the happiest men I know. From the moral point of view it is very difficult! I am not approaching the question from that angle. As you perhaps know, I haven't always been a Christian. I didn't go to religion to make me happy. I always knew a bottle of Port would do that. If you want a religion to make you feel really comfortable,

I certainly don't recommend Christianity." See C.S. Lewis, "Answers to Questions on Christianity," in *God in the Dock: Essays on Theology and Ethics*, Ed. Walter Hooper (Grand Rapids, MI: Eerdmans, 1970), p. 58f.

[197] Tony Blair did was teach a course at Yale entitled, "Globalization and Faith." He did this because he believes these are the most important forces shaping the world today. And fifteen years ago Chinese officials established several hundred centers for the study of Christianity in their country. They did not do this to promote faith in general or Christianity in particular, but because they became convinced that the easiest way to fuel honesty and drive economic prosperity was to persuade people to adopt a Christian worldview

[198] John Adams, and Charles Francis Adams (Ed.), "Letter to the Officers of the First Brigade of the Third Division of the Militia of Massachusetts, 11 October 1798, *The Works of John Adams* (Boston, 1854), vol. 9, pp. 228-229.

[199] I have not heard anyone identify what this new shift entails and am not certain that it is taking place. It seems as though many now realize that postmodernity is not something upon which a life can be built, yet they feel as though there must be something past postmodernity.

[200] See, for instance, Ben R. Crenshaw, "'Shut Up, Bigot!': The Intolerance of Tolerance," *Public Discourse* (Princeton, NJ: The Witherspoon Institute), August 12, 2015.

[201] The other lesser traditional options included Mormonism, Christian Science, Jehovah's Witness, etc.

[202] When I was a junior in college, I had a chance to ask Alvin Toffler a question after a lecture he gave on campus. I wanted to know where religion was headed. His answer surprised me. Toffler said, "People are going to customize. They will pick and choose what they want to believe, essentially creating their own faith."

[203] Sheilaism is also sometimes referred to as "cafeteria religion" for the same reason (its picking and choosing nature). See Robert Bellah, Habits of the Heart: Individualism and Commitment in American Life (Berkeley: University of California Press, 2007).

[204] Globalization, accelerating technology, and changing social norms are disruptive, and during disruptive times many people look for stability by doubling down on their views. We have seen this in the last ten years with Islam, though to be fair, fundamentalist views are also prevalent among Hindus, Christians, and secularists.

[205] In this book I apply the most broad, generic definition of each of the five worldviews, leading to the largest numerical estimates of each. In terms of secularism, bout 15 percent of the people in the United States indicate "no religious preference." This loosely affiliates with secularism, though it is too vague to be an official number. "Nones" refers to those who describe themselves as "spiritual but not religious" and who, like Shelia mentioned earlier, do not participate in "organized religion."

[206] *Touchstone* author Allan Carlson reports the view of self-described secular liberal Eric Kaufmann (*Shall the Religious Inherit the Earth?*): Jews, Christians, and Muslims that allow their scriptures to impact their behavior are "on course to take over the world through demography." A professor of politics at the University of London, Kauffman deconstructs the idea of secular liberalism as advancing, noting "We have embarked on a particularly turbulent phase of history in which the frailty of secular liberalism will grow ever more apparent." Carlson interprets Kaufmann, speaking of liberalism's "ideological exhaustion" (like fascism and communism before it), such that it is "no longer able to inspire self-sacrificing behavior." Carson quotes Kauffman's closing words, and I also think they are worth repeating:

> It will be a century or more before the world completes its demographic transition. There is still too much smoke in the air for us to pick out the peaks and the valleys of the emerging social order. This much seems certain: without ideology [Carlson interprets, "a new secular" ideology] to inspire social cohesion ... [t]he religious shall inherit the earth.

See Allan Carlson, "Saints by Numbers: Will the Religious Inherit the Earth?" *Touchstone*, Vol. 25, No. 2, March/April 2012. The book discussed is Eric Kaufmann, *Shall the Religious Inherit the Earth? Demography and Politics in the Twenty-First Century* (London, Profile Books, 2011).

[207] Buddhists believe that the ultimate goal in life is to achieve enlightenment, which is not found in either luxurious indulgence or self-mortification but in "the middle way." Among the central things taught by Buddha, the "Enlightened One," are the "Four Noble Truths" and the "Eightfold Path." The Four teach that: to live is to suffer, suffering is caused by desire, one can eliminate suffering by eliminating all attachments, and, detachment is achieved by following the Eightfold Path. The Path entails having a right view, intentions, speech, actions, livelihoods, efforts, mindfulness, and concentration. For a point of entrée, see David Llewelyn Snellgrove, "Buddhism (religion)," *Encyclopedia Britannica* (online), July 21, 2016.

[208] For instance, according to Sunni Islamic teaching, a Muslim is one who affirms the Six Articles of Faith (belief in one God, and in angels, the prophets, the Qu'ran, final judgment, and predestination), and practices the Five Pillars (faith, prayer, giving, fasting, and pilgrimage). For the Muslim majority prediction, see Bernard Lewis, *What Went Wrong? The Clash Between Islam and Modernity in the Middle East*, (New York, Harper, 2003).

[209] As Warren notes, 71% of all Christians living in 1900 lived in Europe; by 2000 it had declined to 28%. Conversely, 10% of all Africans in 1900 were Christian, versus over 50% today. Brazil currently sends out more missionaries than Great Britain and Canada combined, and Africa sends more missionaries to Europe than *vice versa.* See Pew Research Center Staff, "A Conversation with Pastor Rick Warren," *PewForum.org*, November 13, 2009. Hence, as far as faith in Christ is concerned, the map has been turned upside down. The joke is that if the U.S. ever gets around to sending a mission to Mars, it will be greeted by Korean missionaries. See also Todd M. Johnson and Researchers at GCTS CSGC, *Christianity in its Global Context*, 1970-2020: Society, Religion, and Mission (South Hamilton, MA: Gordon-Conwell Theological Seminary [GCTS], Center for the Study of Global

Christianity [CSGC], June 2013), pp. 5-10 and *passim.*

[210] Adapted from Mark Sayers, *The Disappearing Church* (Chicago, Moody Publishers, 2016).

[211] Christian Smith's "On 'Moralistic Therapeutic Deism' as U.S. Teenagers' Actual, Tacit, De Facto Religious Faith," gives his label for this misdirected assumption (*Ptsem.edu,* July 21, 2016). Note, the level of commitment in the church varies from a casual, cultural affirmation of the tenets of the faith, to actual, ongoing martyrdom. In light of this, a meaningful, definitive count is difficult to establish. But in keeping with our decision to apply broad, generic definitions of the five worldviews, this puts Christianity at having about two billion current adherents in the three major categories under the Christian umbrella. See Conrad Hackett and Brian J. Grim, *et al.*, "Global Christianity – A Report on the Size and Distribution of the World's Christian Population," *PewForum.org,* December 19, 2011.

[212] Rather than teaching that good people go to heaven, the Bible teaches: **a)** that Jesus is God and always was (i.e., he existed from before time as God in heaven); **b)** that through the incarnation (birth in Bethlehem) he added deity to humanity, showing up as fully God and fully man; **c)** that he became a person in order to love, serve, and teach about God the Father and the coming kingdom, but also **d)** to die to atone (pay) for the sins of the world; and **e)** that he reconciles all of those who receive his offer, and embrace him as Lord.

[213] As a point of entrée, see Brian J. Grim and Mehtab S. Karim, *et al.*, "The Future of the Global Muslim Population, Region: Americas," *PewForum.org,* January 27, 2011.

[214] Regarding the Secularization Thesis: some—e.g., Anthony Gottlieb, former editor of *The Economist*—still believe that is true, and that "[d]espite the flowering of faith in recent times. . . religion will eventually weaken in the developing world." See Daniel Franklin, "Introduction: Meet Megachange," and Anthony Gottlieb, "Believe it or Not," in *Megachange: The World in 2050* [*The Economist,* Vol. 105], D. Franklin and John Andrews, Eds. (New York: John Wiley, 2012), p. xiii, and 79-91.

[215] Though Islamic fundamentalists have generated considerable press during the last ten years, and have obtained control over significant tracts of land I believe their influence will decline over time. I hold this view for several reasons: **a)** there is a revolving door of leadership within the radical Islamic world—the move from the Taliban to Al Queda to Boko Haram to ISIS hints at disunity and an inability to govern. Indeed, at times it seems as though the Islamic State is more of a media conglomerate than anything approaching an organized movement or sovereign state; **b)** though a significant number of Muslims identify with the "jihadist" arm of the faith, a much smaller number (200K) are willing to use violence / engage in terrorism; **c)** moderate Muslims are growing increasingly critical of the radical branch (most of the violence from jihadists is directed at moderate Muslims not Westerners); and **d)** the birth rates among Muslim women are starting to wane. For these reasons (and others) I suspect that over time radical Islam will decline. However, in order for that to happen:

- Muslims need to move away from authoritarian, oppressive political systems.
- Economic development needs to allow people to leave behind lives of desperation.
- Muslims need to understand themselves in ways other than who they are in conflict with others—Jews, Christians, other Muslims (Shi'a vs. Sunni, Sunni vs. Sufi, etc.)

This will be a struggle. International and Middle Eastern conditions remain ripe for continuing Islam's radicalizing. And when conservative views on Islam take over in a country (Iran for instance), the country becomes more radical. Finally, it is likely that Muslim extremists will continue to be a terrorist threat, and a nuclear Iran could always cause havoc in the Middle East. Nevertheless, I think that if you own stock in Radical Islam, it's time to sell.

Chapter Nine, *Next Steps and Final Thoughts*

[216] All things being equal, I'd rather spend time with glass-half-full people than glass-half-empty. But there are two issues. As John McCain noted from his time in a Vietnam prison cell, **a)** those who think things are going to be easy—i.e., we'll be rescued and home by Christmas—are less able to withstand difficulties themselves, and, **b)** they prevent others from preparing. To quote Scott Peck's opening to *A Road Less Traveled,* "Life is difficult." Those who accept this find that it is far less difficult than it might be. Those who expect it to be easy have a very difficult path forward. See M. Scott Peck, *A Road Less Traveled: A New Psychology of Love, Traditional Values and Spiritual Growth* (New York, Simon & Schuster, 2002 [1978]), p. 15.

[217] Verlyn Klinkenborg," The Prophet [Review: *Oil and Honey: The Education of an Unlikely Activist* by Bill McKibben]," *The New York Review of Books,* October 24, 2013.

Mike Woodruff is the senior pastor of Christ Church, a growing, community church with campuses in Lake Forest, Highland Park, and Christ Church Crossroads in Grayslake IL. In addition to founding The Ivy Jungle Network and serving as the President of Scholar Leaders International, Mike has worked as both a college minister and management consultant. As an author he has published over two hundred articles for business and ministry publications and edited or contributed to several books. He holds degrees from DePauw University and Trinity Evangelical Divinity School. Mike and his wife Sheri have three sons, Austin, Benjamin, and Jason. Visit the Christ Church website at www.christchurchil.org and Mike's blog at www.mikewoodruff.org to learn more.

Christ Church is one church in three locations:

Christ Church Crossroads
1350 IL Route 137 Grayslake, IL 60030

Christ Church Highland Park
1713 Green Bay Rd. Highland Park, IL 60035

Christ Church Lake Forest
100 N. Waukegan Rd. Lake Forest, IL 60045

For more information on service times
call 847.234.1001 or visit www.christchurchil.org